H. Tsubaki, K. Nishina, S. Yamada (Eds.)

The Grammar of Technology Development

H. Tsubaki, K. Nishina, S. Yamada (Eds.)

The Grammar of Technology Development

With 64 Figures

Springer

Hiroe Tsubaki, Ph.D.
Professor
Graduate School of Business Sciences
University of Tsukuba
3-29-1 Otsuka, Bunkyo-ku,
Tokyo 112-0012, Japan

Ken Nishina, Ph.D.
Professor
Nagoya Institute of Technology
Gokiso-cho, Showa-ku, Nagoya,
Aichi 466-8555, Japan

Shu Yamada, Ph.D.
Professor
Graduate School of Business Sciences
University of Tsukuba
3-29-1 Otsuka, Bunkyo-ku,
Tokyo 112-0012, Japan

ISBN 978-4-431-99836-5 e-ISBN 978-4-431-75232-5

Springer is a part of Springer Science+Business Media
springer.com

Printed in Japan

Printed on acid-free paper

This book is dedicated to Mr. Akira Takahashi who passed away during the editing of this book.

Preface

This book includes the keynote lecture and fourteen selected papers that describe a general guideline and supporting concepts and tools for conceiving technology development as a grammar. Recent advances in scientific and engineering fields call for new disciplines, tools, and concepts. For example, advances in computer simulation require new approaches to statistical techniques to utilize computer simulation efficiently for technology development. The papers collected in this book focus on such new approaches based on these practical requirements. The editors are confident this collection will contribute to the acceleration of technology development through the application of the grammar of technology presented here.

The title of this book is influenced by Karl Pearson's book *The Grammar of Science*, published in 1892, which brought him recognition as a giant and pioneer of statistics. His book introduced a grammar of science with a description of the roles of statistical treatments. While science at times has been misunderstood as not being amenable to a standardized approach, one of the contributions of Pearson's book was that it offered a standardized approach to science. As his book demonstrated, behind the great innovations of science, there exists a universal approach.

One of the present volume's editors, Hiroe Tsubaki, has insisted on the necessity of developing a new general approach to technology development based on scientific and engineering advancements over time. This direction is gradually gaining attention among many practitioners and academicians. In other words, researchers and academicians are drawn to develop a grammar of technology development as a standardized approach. Fortunately, the project to develop a grammar of technology development has been supported by many organizations, including the Japanese Ministry of Education, the Japanese Society for Quality Control (JSQC), the University of Tsukuba, and others.

In particular, the research proposal by Hiroe Tsubaki to develop the grammar of technology development qualified for a Grant-in-Aid for Scientific Research from the Ministry of Education, Culture, Science and Technology, Japan. Such interest has accelerated the development of the grammar of technology development, and was the impetus for a meeting around a research project on May 15, 2004, at the Nagoya Institute of Technology. At that meeting, Akira Takahashi, president of JSQC 2003–2004, gave the keynote address. That meeting can be regarded as the trigger for this ambitious project.

The following year, a workshop titled, "The Grammar of Technology Development," was held on January 15–16, 2005, at the University of Tsukuba. It was a great opportunity to develop concepts, tools, and a grammar from a variety of interdisciplinary viewpoints. After the workshop, we invited the presenters to submit their papers in order to take the first step toward the development of the grammar of technology development based on the discussions at the workshop. Consequently, we are happy to include nine papers along with the keynote lecture by Akira Takahashi from this inaugural workshop. Five additional papers were included after being reviewed by at least two experts. The papers are classified into the following three parts:

Part I Systematic Approaches to Technology Development
Part II DE^2: Design of Experiments in Digital Engineering
Part III Statistical Methods for Technology Development

The papers in Part 1 relate to a general and systematic approach to technology development. The first paper sets out the fundamental strategy underlying the grammar of technology development. There are various viewpoints discussed regarding the systematic approach to technology development, such as informed systems, quality function development for the environment, and communication management. The final paper in this section presents a case study of highly advanced technology development in a production system.

The papers in Part 2 discuss the applications of the design for experiments in the digital engineering field. A grammar for the design of experiments in computer simulation is introduced at the beginning of this section. Applications of certain advanced techniques in the design of experiments are discussed, such as uniform design and the response surface method. This section also includes a discussion of the technological trends and applications of the design of experiments in practice, particularly in the manufacturing sector.

In Part 3, statistical methods for technology development, such as hybrid simulation, and a sampler from discrete Dirichlet distribution are considered. In addition, a paper on the application of the maximum likelihood method to communication systems is included. The final paper in this section shows an application of the statistical method to visualize melody sequences.

These collected papers should provide readers with advanced knowledge of technology development.

The editors would like to express their gratitude for the Grant-in-Aid for Scientific Research 16200021 of the Ministry of Education, Culture, Science and Technology, Japan, where Hiroe Tsubaki is a representative. Furthermore, we thank the Japanese Society for Quality Control and the University of Tsukuba for their support.

During the editing of this book, Mr. Akira Takahashi, a great supporter of our activities, sadly passed away on August 14, 2006. His support of our endeavors was manifold. For example, while he was president of JSQC, he supported the formation of a research group on simulation and statistical quality control to investigate effective approaches to quality management, and he had provided much valuable feedback on the work. All of the editors and some authors of this book were members of that research group. His keynote speech at the kick-off meeting at the Nagoya Institute of Technology further attests to his support of our activities.

We express our sincere condolences upon his death and offer this book as a tribute to his foresight and spirit.

Hiroe Tsubaki, Ken Nishina, and Shu Yamada
Editors

Contents

Fusion of Digital Engineering and Statistical Approach

Akira Takahashi

DENSO CORPORATION, Chairman
The Japanese Society for Quality Control, Advisor

1 Introduction

Thank you, Dr. Nishina, for that introduction. It is a real pleasure to speak to you all today!

I'd like to start with a brief account of how I came to be invited to give the keynote lecture.

I'm afraid that someone as unfit as I am for the academic world being hauled up to give this kind of address is a perfect illustration of the proverb: "Who spits against heaven, spits in his own face."

Actually, I have been conscious of a problem in engineering for some time now. Most of the time that we spend, as engineers in industry, is on the inductive approach, using the inductive approach to find solutions. One of the applications of this is statistical quality control (SQC), that is to say, statistical approach. However, once having solved particular problems, they do not follow through and find general solutions. I think that is because this step would require us to add the deductive approach to our thinking. Meanwhile, computer-aided engineering (CAE) is becoming a more and more important tool. We have been using it more and more every day. CAE incorporates both the deductive and the inductive approaches. Another very important theme to be dealt with in industrial research is how to handle dispersion in data. As a matter of fact, I have already asked the authorities in The Japanese Society for Quality Control (JSQC) to consider this as one of their most important topics. Dr. Nishina has already described all that in detail; Dr. Nishina, Dr. Tsubaki and many other researchers will join in and participate in the project of Transdisciplinary Federation of Science and Technology. JSQC will organize the research, with me as the chairperson. I had never dreamed that any of that work would also be assigned to me.

So, this conference has begun with spitting against heaven, and now, we are in the keynote lecture.

2 The consciousness of the outsider

Next, I'd like to explain how I came to realize the points I just raised by describing my experiences.

I have been working in industry for a long time. I was completely ignorant about digital engineering and SQC. What I had done was only go through a basic course in SQC. I can't talk to you about mathematical principles or procedures — I can't be of any help to you there. But if you stay with me for a while, I think you will understand how our theme for today originated in the history of the industrial work environment.

Figure 1 shows a timeline of how digital engineering and SQC became part of my work. Actually, my first job was a production engineer at Toyota Motor Corporation. That was from 1965 to 1970. NC machines had just come out at the time. We began using NC machines for diesinking. We programmed the machines using a tape to shape the stamping die. In turn, this was replaced by direct numerical control or computer numerical control, where using a computer to control the machine directly made the tape unnecessary. I worked very hard in the revolution in manufacturing processes. Up to that time, the stamping die for the car was made to imitate the car model in advance. The presses had been used to design the models, but with NC, we were using what we call "digital thinking" in Japan now. Thus, the transition to the digital era started in the production departments.

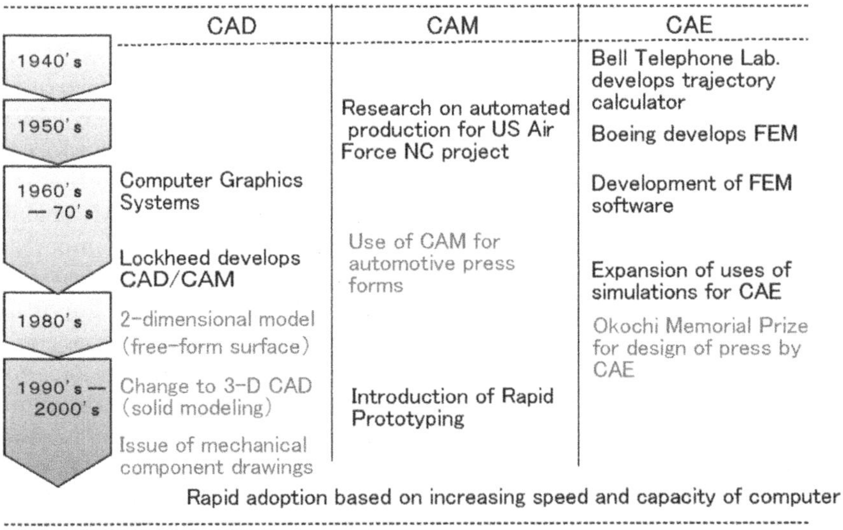

Fig. 1. Digital Engineering in my job

An automobile body is composed of free-form surfaces. The designers express those curves with line drawings. I don't know if you all know this, but they draw contour curves for the entire car. They show this with lines on a two-dimensional piece of paper. It is a very difficult task to create an image of a free-form surface with a line graph. We heard of people who, faced with this job for the first time in their careers in stamping die operations, went neurotic, or lost the hair on their heads. The job demanded special skills. But then, the course of digitization continued. Eventually, 3-D solid modeling came on line and became convenient to use. The need for a way to express the free curves in stamping die drove the body design departments to incorporate digital engineering.

So digital engineering has made strides into the workplace thanks to the free curves on automobiles. The engine, transmission and other purely mechanical components are built up of geometrical shapes, so there was not so much need for solid modeling. Finally, there has been great progress in digitizing drawing of mechanical components in recent years, and CAE is coming into wide use. And, while previously only the third angle projection method used to be acceptable for approved drawings of components, just very recently, it has become acceptable to submit CAD data.

Figure 2 shows a job I worked on, my first in CAE. I am rather proud of this one. I won the Okochi Memorial Prize for this in 1985. As I mentioned earlier, it is very difficult to express where different curves intersect together on a stamping form, especially on a rear quarter panel. No small number of strains occur while a steel sheet is straining. To correct the form, there was a long process of trial and error to complete the shape of cars. That was just one of the bottlenecks in preparing to produce cars. So CAE was developed in the hope of breaking through this. The CAE we had then was nothing like the CAE we have now. Even if calculation by FEM (finite element method) on one of the then-available supercomputers, it would take days to get a solution. It was unusable for most design work.

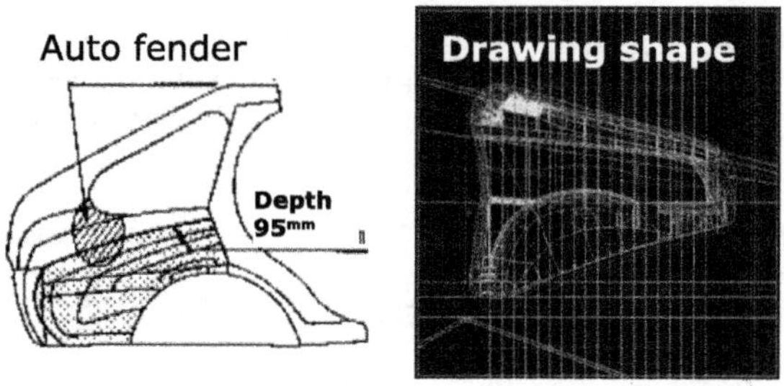

Fig. 2. Curved automotive body is difficult to represent

So, at that time, I had a CAE system run a calculation of a surface like in Figure 3. It predicted in what directions and in what fashion the steel sheet would initially spread out. If there were many discrepancies among the elongations, it meant that there would be strains in those locations. The point was to determine how to design the curve on the die face so as to eliminate strain. This CAE did that calculation.

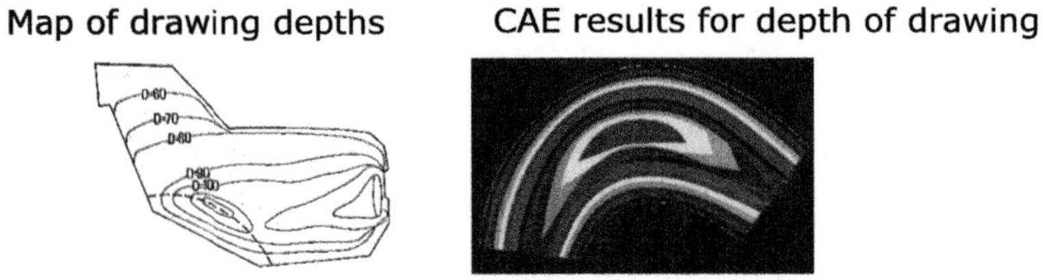

Fig. 3. Analysis of deformation in the rear quarter panel

Actually, when we laid a steel sheet on a stamping die, it does not lie flat. The first process in forming is to create this curved surface, and then to lay the sheet on it. If the sheet was paper, it would become crumpled, but steel is ductile, so we lay it down on the curves of the form and gradually push it into place with punches. But, in some places, strains may begin to occur in one side. In order to prevent this, I adopted a new interactive method for designing the curves on the die faces. I developed a hypothesis for calculating changes in elongation, changes in the contour line heights with punch impact and other parameters. This was one of my success stories.

Recently new ideas have become widely accepted for shortening lead times. For example, there is "front loading", which means that design problems are anticipated and solved as far in advance as possible. There is also concurrent engineering, and the "V-comm" approach. I will discuss V-comm later. Obviously, digital engineering is absolutely necessary in order to push forward with shortening the lead time, the globalized synchronization of the development and the production. Figure 4 shows this in terms of the lead time. What used to be consecutive steps of design, mock-up, evaluation, output of drawings, process evaluation, preparation for production, etc., are now concurrent efforts. Many processes go on at the same time. When I was at Toyota, we released a car called the bB, a brother car of theVitz/Yaris. The bB used the

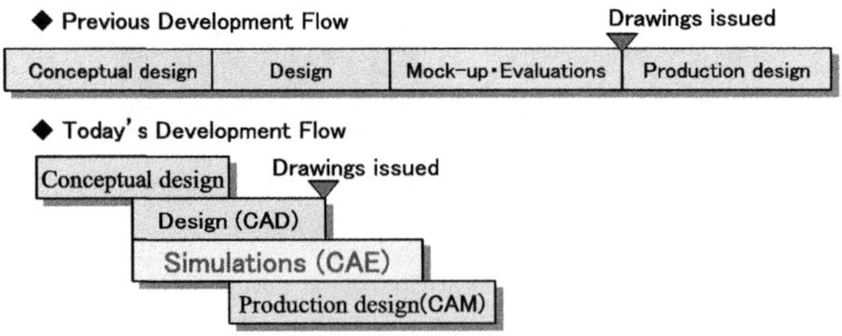

Fig. 4. Development flows before and after the change-over to digital engineering

Vitz's underbody. Only its upper body was new, so it was developed without mock-ups. We set ourselves the challenge of developing that car in a year, using digital engineering. In the end, it took 13 months. Figure 5 compares lead times in the automotive industry around 1998 and now. But concurrent engineering is being developed in other industries as well, not just automobiles.

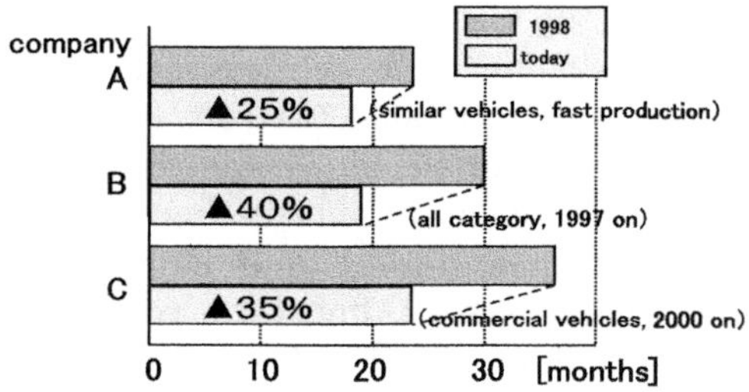

Fig. 5. Development periods around 1998 and today

Well then, let me describe V-comm. This is a means of communication in which we place a screen big enough to display a life-size car in every design department in the world.

I mentioned the front-loading before. This is being used to quickly solve problems like those shown in Figure 6., layout interference, productivity and appearance. In the factory, we used to find about 10,000 cases in each model requiring changes. That has been cut to one twentieth, and now, we are reducing it even further. I mentioned the lead time. We used to go through several design generations with the die face. Now, we are shooting in one generation. I think Toyota is doing their mock-ups in an actual model off the line now.

Fig. 6. Development environment with V-comm

They used to make first, second, third and even more mock-ups, as many as necessary, but now, they just need one cycle.

Thus, digital engineering has made tremendous progress. It's contributing to efficiency. There's another side, however, that I am really concerned about, as a top management of technology. The engineers use this tool without thinking about the meaning of the job they're doing. The situation has arisen where they can get answers by using these tools, and they do not need to know anything about the meaning of the tasks in order to get their job done. Also, the level of technical expertise is falling. This is true outside technological fields as well. It's also true in education, probably not just at universities, but at the elementary and junior high school levels. They're just looking for the right answers. This is not just a problem of the teachers. We, top management, have this problem too. We demand results, and don't make our subordinates think carefully through the processes. Almost everyone is doing this.

I've been very concerned about this. I've been ringing the alarm bell about it, and I talk to engineers in the company about 2 things all the time. What I say to them is, "Return to the worksite and see the actual product", and "Realize the importance of measuring and visualizing".

When someone works with a black box using digital engineering, not only do they stop touching or looking at the actual product, often they will fail to recognize the product even when they are looking right at it. Maybe they do not bother taking measurements either. They stop going to the factory where a product is making and spends more and more time in front of their monitor. I think this is unacceptable. It is essential for engineers to nurture their intuition for the product, in order to broaden their technical abilities. To nurture his intuition, they have to spend time in the factory, looking at the product, training all five senses. Using the five senses at the computer does not develop them at all.

Let me discuss the other issue, measurement and visualization, a little bit. In order to resolve problems during research and development, the most important practices for understanding phenomena are measurement and visualization. "Visualization" is a category among measurements. When I say it's important, I mean that Toyota has set up a measurements technology section and they're studying visualization. Denso has a laboratory called Nippon Soken, Inc., full of academically talented researchers. Denso is supporting them with the goal of extending its own measurement technologies.

For ordinary companies, however, these technologies are incidental. They don't consider them very important. Many companies force such work on their subsidiaries, or cut back on them as a cost-cutting strategy. They cut their own throats when they do that. I consider this thinking to be a very serious problem, and I bring it up with people all the time.

3 My expectations of SQC activities in industry

We have an SQC study group in my company and I'm often asked to talk to them. The engineers in industry nearly always use the inductive approach when they have to solve a problem. They gather a lot of field data and can draw on a good fund of work experience, but when I see them doing the same jobs over and over at every model change, I get the feeling that all they are getting is particular solutions. It is fine if they continue using the inductive approach, but I want them to make some effort to go beyond their immediate issue and come up with a general solution to such issues. I discuss that in my speeches to the SQC study group.

A lot of engineers don't seem to have any consciousness of what kind of approach they are taking at all. They just do what their superiors tell them to do, over and over. Almost all of them follow the inductive approach, but what they need to do is transform some of that effort into deduction, and develop new methods that will prevent problems. This is about creating new know-how in the company. I mean, engineers are making correct distinctions between when to use inductive logic and when to use deductive logic. It is important what kind of consciousness they have as they use CAE and digital engineering. Too many engineers are satisfied with solutions for their own special cases, and are not seeking general solutions.

I mentioned what I consider the danger of black boxes before. Another thing that I'm very concerned about is engineers getting too comfortable with digital engineering. They even start to lose their ability to think deductively. Digital engineering is used in probably 100% of the industrial companies in Japan. I wonder it is possible somehow to apply some statistical approaches? I have some prescriptions based on my experience.

To summarize this a little more, we get Figure 7. We know some of the physical principles involved. We use those deductively to create simulations. We then use the simulation results inductively to create empirical principles. Those are reflected in the physical principles. The physical model is revised and a CAE model is developed to perform the calculations for the physical model. The CAE model is then run and its results are examined closely to verify their consistency with the empirical principles. I think that repeating this kind of cycle is the ideal way to create these simulations.

I will discuss later what I got smattering of American situations now. There are two themes, verification and validation. These are called "V & V". Essentially, verification means taking a very energetic approach, studying how to deductively create a physical model. In this field, they model things that are very difficult to model. They start out with a small one, at some miniaturized scale, and the simplified version is gradually expanded until they ultimately have the model they want. I interpreted "verification" as meaning that whole process. "Validation" means the process of checking whether the simulation and the empirical principles, or the real product, actually agree with each other. Anyway the Americans are doing a lot of research in V & V now. The

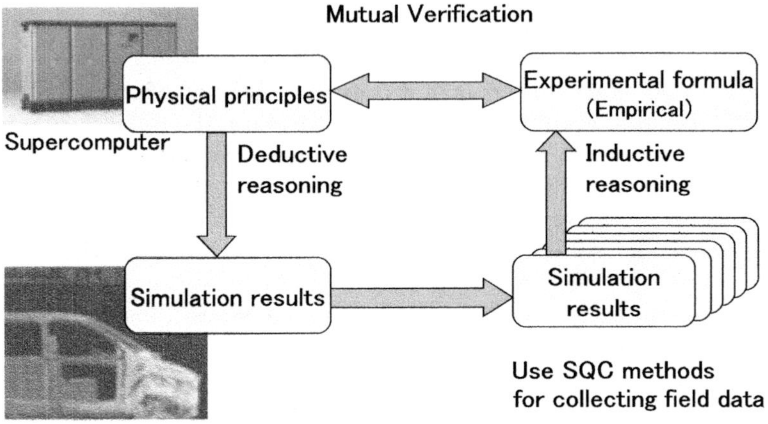

Fig. 7. Combination of deductive and inductive approaches

above is the reason why I started to discuss the fusion of CAE and statistical approach, and also the intention I requested authorities in JSQC.

4 How is CAE used in industry now?

I'm going to get away from the main topic for a few minutes and discuss some specific problems in my own company. Figure 8 shows the answers I got when I asked the engineers to draw maps of their jobs in the company. There are a lot of solution methods here, but the left side is fields for which we already have physical models, and the right side is fields with no physical models yet.

This is too complex to understand for most of you, so I re-arranged the topics in Figure 9. The center represents fields for which our models have acceptable prediction accuracy. The next ring out from the middle is fields where we get medium accuracy, and the outer ring is fields where we can make no predictions using any models yet. This is a concrete condition of my company.

I said that 100% of companies are using CAE before; they have applied it as a part of tasks in various processes. Figure 10 Part A shows situations where prediction accuracy is sufficient. These are situations where the simulations, involving simple equations match well with measurements in actual experiments like the generalized Hooke's law. At this level, the components designed on CAE can be used for production, once they have passed the final experiments. Part B shows fields which can't be considered "company know-how" yet, until we have more experience with them, and another CAE cycle needs to be done so that we can refine the coefficients. Part C means that we have no physical model. CAE cannot be used at all. We can classify the fields into these 3 classes.

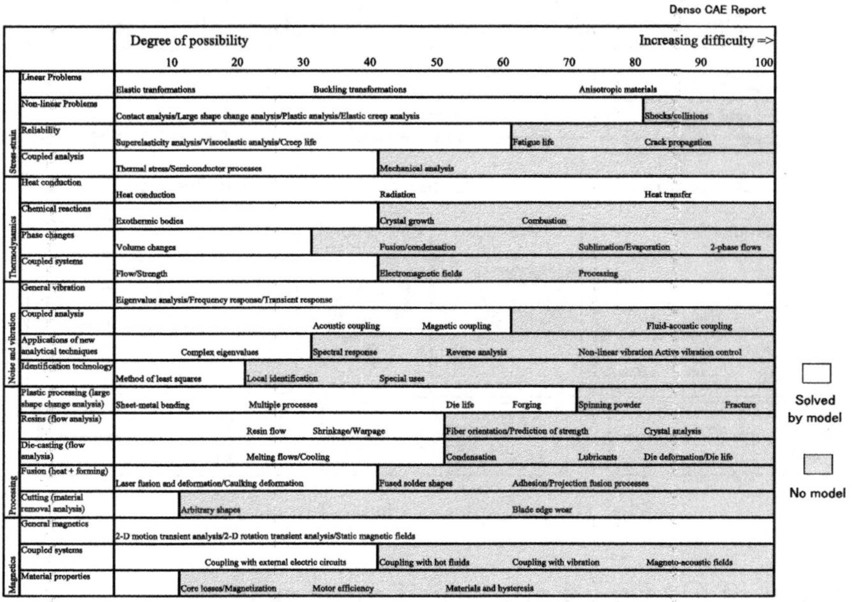

Fig. 8. Map of CAE and physical models (1)

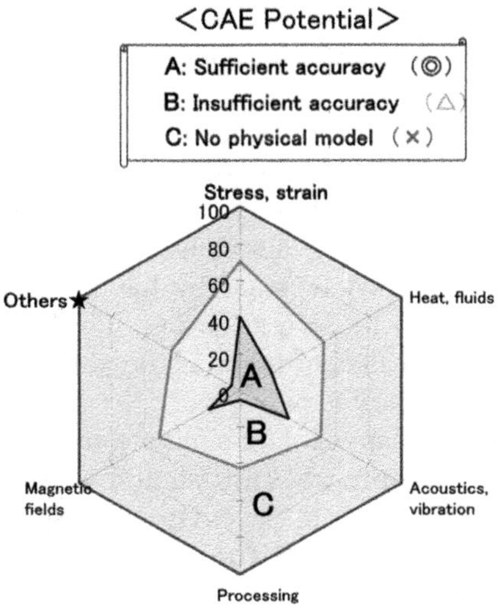

< Physical models >

	A	B	C
Prediction	Possible		Impossible
Accuracy	◎	△	×
Stress, Strain	・Elasticity	・Plasticity ・Creep	・Shock ・Mechanical coupling ・Cracking, life
Heat, fluid s	・Heat insulation	・Conduction ・Flow	・Fusion・Condensation ・Sublimation・Evaporation ・2-phase flow
Acoustics, vibration	・Resonance	・Frequency response	・Flow noise ・Spectral response
Processing		・Plastic processing ・Forming	・Welding ・Cutting ・Die life
Magnetic field	・Static magnetic fields	・Dynamic magnetic fields	・Magnetic noise ・Magnetization ・Coupling
Others★		・Optics ・Circuits	・Irregular reflection ・EMI noise

Fig. 9. Map of CAE and physical models (2)

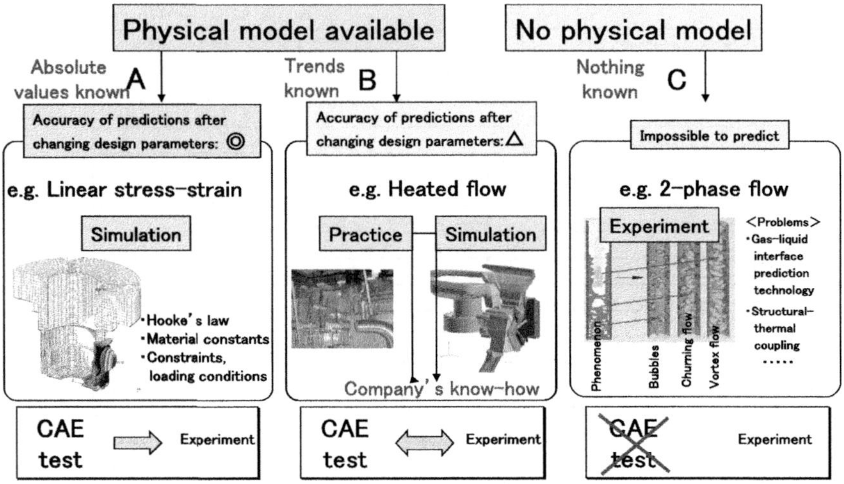

Fig. 10. Examples of physical models

5 Fusion of CAE and statistical approach

From now, I'm going to turn to my own proposals. Today is the "kick off" meeting of Transdisciplinary Federation of Science and Technology. I wondered what steps the study meeting would take in investigations in this event. I thought along 3 lines in my own ideas.

First of all, I think that Step 1 needs to be a thorough investigation of the present situation. The survey will help put us on the same page in our consciousness of the problems. This is Transdisciplinary Federation of Science and Technology, after all, so each of our member organizations has its own specialized field of expertise. I think that we need to start by having each member organization present the main issues in its own situation. Then, we can form a shared vision of the problems, and decide together on different organizations' roles and targets for solving the problems.

The second step would be to reproduce the dispersion in simulations. This is absolutely essential for applying statistical approach. How should we deal with this dispersion that comes from the uncertainties in the environment? Or, to put it in the terms of Figure 11, how should we deal with the various exterior disturbances for which we can set no parameters? To introduce the dispersion into factors for which the parameters are already set, we can incorporate normal random numbers into the parameters, but I think the problem really is, how do we incorporate parameters that we cannot set? Perhaps some genius will give us a general method for incorporating everything some day. For the time being, I put my trust in a consecutive process of enlarging the model, step by step.

That brings me to Step 3! And, of course, this step is to tie all the results together into a single methodology.

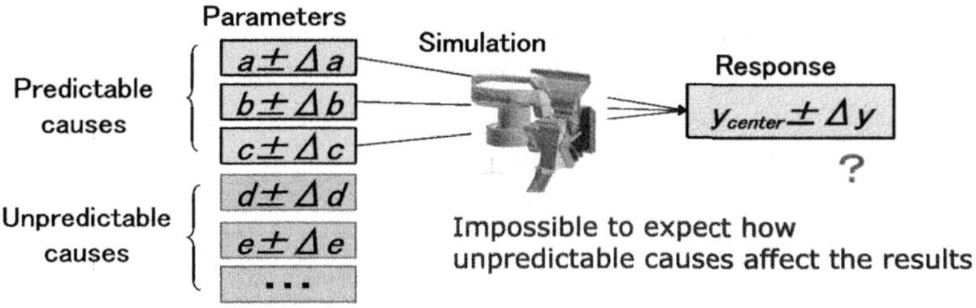

Fig. 11. Parameter settings and dispersion reproducing

6 Conclusion

Before I conclude, I would like to discuss another topic, the current situation in the U.S. The national researchers, especially the military and NASA researchers, have been publishing a lot of experimental data since the space shuttle accident. This has made it easier to conduct V & V of numerical models. A Mach 7 aircraft was created entirely in simulation without any mock-ups. It was flight-tested in March of this year (2004).

V & V means "verification and validation", as I explained before. According to my understanding, in "verification", you start with a primitive model, then improve it gradually with finer and finer details until you have a model you can use. "Validation" means checking whether the results from the model actually match the results in the real product.

I talked with Dr. Noboru Kikuchi of the University of Michigan about the subject of a fusion of CAE and statistical approach. We talked about starting research for JSQC. Dr. Kikuchi studies CAE and I thought he might be able to give me some advice. When I described it to him, he said that they are already studying it in the U.S. Dr. Kikuchi is doing it himself. He said that this is extremely important and JSQC or the Transdisciplinary Federation of Science and Technology should definitely get started on it. So I am encouraging you to do it. The Americans have already been focusing on this topic.

The "hottest" topic now is noise. This problem is a combination of the uncertainties in an acoustic flow field and the dispersion in the surrounding objects, and how to analyze these factors. Another topic after this is biomedical fields. The military is said to have a huge volume of data on this, and researchers are using it. I understand that the medical research group of The Union of Japanese Scientists and Engineers are studying V & V in their symposiums and study group meetings. The Japanese and Americans are apparently doing the same thing.

I borrowed a lot of content from Dr. Kikuchi, but it is essential to do this kind of research for the advancement of science and technology in Japan. I hope that you will make great efforts towards this. This will bring refinements in CAE and enable creation of new models, such as in the fields I mentioned before where, as yet, no models have been developed. And I hope that you will demonstrate how to create the models that we need the most. As I mentioned before, in order for this to be really useful, measurement technology must also be upgraded. I'm sure all of you agree on that! We must team up with organizations that actually make things.

Finally, I'd like to touch on American V & V. I think it's pretty easy to agree with the V & V approach, but I hope to see much academic work in this area. I'm sure companies will take this approach in order to meet their own needs, but, just speaking for myself, I'd be very happy to see academic societies on this topic as well.

There are various difficult problems in use that I have not touched. For that, I pass the baton on to the specialists. Thank you very much for your kind attention.

(Revised and translated the keynote lecture at the workshop "the subject of quality control in digital engineering era", May 15, 2004, Nagoya Institute of Technology, Japan)

Part I

Systematic Approaches to Technology Development

The Grammar of Technology Development
— Needs for A New Paradigm of Statistical Thinking —

Hiroe Tsubaki

Graduate School of Business Sciences
University of Tsukuba, Tokyo JAPAN
tsubaki@mbaib.gsbs.tsukuba.ac.jp

Summary. Grammar of technology development is a trans-disciplinary description of common approaches to well-controlled technology developments in which the most effective method for development is systematically selected. Here technology development involves the following four sequential activities for both a real society and the corresponding virtual society using appropriate engineering models:

1) Value selection of targets by defining the expected recognized quality elements.
2) Translation of the recognized quality elements occurring in societies into functional quality elements that designers and engineers can specify concrete parameters in their engineering models.
3) Optimization of design parameters of the engineering models to ascertain their usability.
4) Value injection into the real society to harmonize realized functional qualities and corresponding recognized quality.

Each activity above may be partly supported by classical statistical thinking established by Karl Pearson's grammar of descriptive sciences in the late 19th century, however, a new statistical thinking should be introduced for evaluating results not derived from observation of a real community but from a simulation on a virtual society. Geninchi Taguchi and his colleagues have already developed such methodology for optimization activities implicitly based on such new thinking for technology development. The paper describes an aspect of new statistical thinking for technology development through reviews and interpretation of Taguchi's contribution to the design of experiments.

1 General Introduction

A specialist in the grammar of technology development should be a leader in improvement. So what is the grammar of technology development? And how is it possible to attain systematic technology development by applying such a grammar? One of the objectives of this book is to suggest answers to the above questions and to clarify new roles of statistical thinking in technology development.

I myself currently think that the grammar is a trans-disciplinary model for describing a general process of well-controlled technology development by which we scientifically select the most effective approaches to success. As an initial working hypothesis I propose technology development as involving the following 4 activities applied to the 3 fields as shown in Fig. 1.

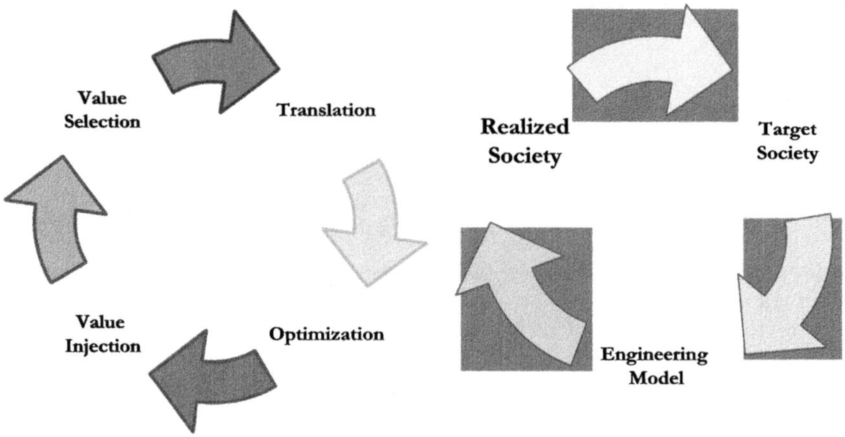

Fig. 1. Modeling the Technology Development Process

The 3 fields:

Field 1: Real or realized society where users of the product live.
Field 2: Virtual target society affected by the designed technology.
Field 3: Engineering models describing the interaction between the product and the society for the purpose of research and development.

The 4 steps:

S STEP: Value Selection

Objectives: Selection of values with targets by defining the expected recognized quality elements or the expected voice of the customer (VOC), in other words, design of the target society based on the scientific recognition of real society either now or in the future.

Methods: Predicting and analyzing the difference of the user's performance between real society and the virtual target society created by the designed technology to clarify the target society. Exploratory statistical tools will be useful for residual analysis of statistical prediction models or clustering techniques in the initial stage of the S step, but some confirmatory approach, such as conjoint analysis, may be also necessary for the final value selection.

T STEP: Translation

Objectives: Translation of the recognized quality elements from real societies into functional quality elements that will allow designers and engineers of the product to specify concrete parameters in their engineering models.

Methods: Clarifying systems to attain the requirements of the expected society. Quality functional deployment (QFD) and cause and effect diagrams will be helpful analysis tools for the T step.

O STEP: Optimization

Objectives: Attainment of usability in terms of the specifications defined in step 2 by optimizing the design parameters of the engineering models.

Methods: Designing and implementing real experiments or simulations to evaluate the best performing parameters of the engineering model against the variation of the noise expected in the target society.

I STEP: Value Injection

Objectives: Achieving consistency between the realized functional qualities and corresponding recognized qualities in the real society.

Methods: Communication and information management to inform the users of the value of the developed products or functions.

I expect the each step should have its own grammar supported by both classical and modern statistical thinking. This book describes the roles of chance management or knowledge management in the S and V steps, and a new style of QFD in the T step and experimental designs for the numerical simulation in the O step. It also gives several real-life expamples of statistical thinking applied to new technology developments.

In the next section, I briefly review the history of the grammar of science and the needs for new statistical thinking in the optimization step of designing new technology.

2 A Review: Role of Classical Statistical Thinking in the Grammar of Science

Since the late 19th century statistical methods have been developed for supporting cognitive processes in modern sciences by positivists affected by Darwinism, such as F. Galton and K. Pearson. Several useful concepts and tools have also been established for knowledge discovery or optimizing a new technology, including "process control" by W. Shewhart (1931), "design of experiments for researchers" established by R.A. Fisher (1935) and "design of experiments for technology development" which was developed by G.E.P. Box et al. (1978), G. Taguchi (1986) and other Japanese researchers of industrial experiments as T. Okuno, T. Haga and Y. Washio since the early 1950s.

K. Pearson established the modern statistical thinking with his "Grammar of Science." The 1st edition of "The Grammar of Science" was published in

1892 shortly after the author had given lectures to the citizens of London from 1891 to 1892. He stated that "Man gives a law to Nature" in this memorial lecture series by suggesting a concrete and systematic way to construct a science or a scientific law. His ideas might originate from F. Galton (1883), who established statistical science and who defined its objective as follows:

> "The object of statistical science is to discover methods of condensing information concerning large groups of allied facts into brief and compendious expressions suitable for discussion."

However it should be pointed out that Pearson himself introduced a systematic way for constructing "Science" and its fundamental concepts as follows:

1) Empirical or probabilistic interpretation of cause and effect.
2) Statistical description of a scientific law by contingency and correlation.
3) Classification of facts into sub groups to improve the performance of the law.

He also developed basic statistical methods and concepts such as the histogram, the standard deviation and the regression function to help construct the science he defined in his grammar. For instance his correlation coefficient is regarded as a performance measure of an observed scientific law that is independent of the origin and scale of the measurements and is thought to have accelerated the development of statistical sciences such as biometrics, psychometrics, and econometrics and technometrics in the early 20th century.

In Pearson's statistical science, phenomena are described by a statistical model given by

$$\mathbf{Y} = f(\mathbf{x}) + \varepsilon,$$

where $\mathbf{Y}$ and $\mathbf{x}$ are q and p dimensional output and input vectors, respectively, and the function f is classified as a scientific law for the case when f is simple enough to interpret and the error ε is considered sufficiently small to apply the formula for control or prediction. Traditionally a simple model means a linear model

$$\mathbf{Y} = \mathbf{B}\mathbf{x} + \varepsilon,$$

or a log-linear model, such as

$$\log \mathbf{Y} = \mathbf{A}(\log \mathbf{x}) + \varepsilon,$$

with the constraint $\mathbf{A1} = \mathbf{1}$ as a consequence of the property of "a constant returns to scale", and the dynamic characteristics of a composite system can be easily estimated if its subsystems are linear. In traditional statistical methods the error ε has further been assumed to follow some probability distribution by using randomization techniques if necessary.

3 A New Paradigm of Statistical Thinking for Technology Development

F. Galton (1883) proposed that the aim of descriptive science or science for recognition is to achieve a consensus among researchers and this consensus is frequently termed an approved scientific law based on a necessary and sufficient classification of the observed facts and an evaluation of their correlation (Pearson, 1892). By contrast, the aim of technological development should be to extend an observed scientific law to conform to a technological purpose, which is based on a certain kind of value selection. Therefore we could claim "Man can improve the law for his own purposes." In fact the classical design of experiments described by Fisher (1935) may treat improvement of the concerning characteristics and proposed useful methods, such as randomization, replication and blocking to generalize the experimental findings. However, the above-mentioned researchers did not focus on extending an applicable scientific law to research and development.

To achieve this goal, Taguchi suggested incorporating designable parameters into the scientific laws in the following manner

$$\mathbf{Y} = f(\mathbf{x} \mid \mathbf{C}, \mathbf{N}) \tag{1}$$

where $\mathbf{C}$ is a c dimensional vector of controllable factors, the levels of which can be controlled and optimized both by performing experiments in the laboratory and by using in production or the market. Also $\mathbf{N}$ is an n dimensional vector of noise factors, the levels of which can be controlled only in experiments testing the engineering models in the laboratory: they are considered random variables in the target society with the production process or market.

Taguchi introduced a desirable technical target for the linearity of the relationship between the inputs $\mathbf{x}$ and outputs $\mathbf{Y}$, and defined the "ideal function" as

$$\exists \mathbf{C} \in \Gamma,\ \forall \mathbf{N} \in \mathrm{N},\ \forall \mathbf{x} \in X, \quad \mathbf{Y} \approx \mathbf{Bx}, \tag{2}$$

where Γ, N and X are the feasible design space, the expected noise space and the setting signal space, respectively. The input $\mathbf{x}$ may be a kind of controllable factor, however Taguchi called it a vector of signal factors which designers can use to attain an approximate linear relationship between the inputs and outputs given in (2).

4 Necessary Fusion of the Taguchi and Model-based approaches

Many articles such as Wu and Hamada (2000) have been published attempting to interpret the Taguchi methods using appropriate statistical models. However the standard Taguchi method optimizes the performance of a system by measuring directly the discrepancy between the ideal and real function, without identifying and fitting a model to observations of the system. This is in contrast to traditional statistical methods, such as the response surface method, which interpret phenomena only by using appropriate modeling.

But Taguchi also might have been using the model implicitly and his ideal function given in (2) should be considered a tool to measure or optimize the performance of engineering models. Therefore traditional mathematical modeling, can be used for the optimization provided it includes the concepts of controllable or noise factors as shown in (1).

To verify this I consider the scalar output of the 2nd order approximation, or the quadratic response surface approximation by G.E.P. Box, of the system given in (1) as

$$
\begin{aligned}
Y &\approx Y_0+\mathbf{B}_{\mathbf{x}}^{\mathbf{T}}\mathbf{x}+\mathbf{B}_{\mathbf{C}}^{\mathbf{T}}\mathbf{C}+\mathbf{B}_{\mathbf{N}}^{\mathbf{T}}\mathbf{N}+\frac{1}{2}(\mathbf{x}^{\mathbf{T}},\mathbf{C}^{\mathbf{T}},\mathbf{N}^{\mathbf{T}})\begin{pmatrix}\mathbf{H}_{\mathbf{xx}} & \mathbf{H}_{\mathbf{Cx}}^{\mathbf{T}} & \mathbf{H}_{\mathbf{Nx}}^{\mathbf{T}}\\ \mathbf{H}_{\mathbf{Cx}} & \mathbf{H}_{\mathbf{CC}} & \mathbf{H}_{\mathbf{NC}}^{\mathbf{T}}\\ \mathbf{H}_{\mathbf{Nx}} & \mathbf{H}_{\mathbf{NC}} & \mathbf{H}_{\mathbf{NN}}\end{pmatrix}\begin{pmatrix}\mathbf{x}\\ \mathbf{C}\\ \mathbf{N}\end{pmatrix}\\
&= Y_0+\mathbf{B}_{\mathbf{C}}^{\mathbf{T}}\mathbf{C}+\mathbf{B}_{\mathbf{N}}^{\mathbf{T}}\mathbf{N}+\frac{1}{2}(\mathbf{C}^{\mathbf{T}},\mathbf{N}^{\mathbf{T}})\begin{pmatrix}\mathbf{H}_{\mathbf{CC}} & \mathbf{H}_{\mathbf{NC}}^{\mathbf{T}}\\ \mathbf{H}_{\mathbf{NC}} & \mathbf{H}_{\mathbf{NN}}\end{pmatrix}\begin{pmatrix}\mathbf{C}\\ \mathbf{N}\end{pmatrix}\\
&\quad+\left(\mathbf{B}_{\mathbf{x}}^{\mathbf{T}}+\mathbf{x}^{\mathbf{T}}\mathbf{H}_{\mathbf{xx}}+\mathbf{C}^{\mathbf{T}}\mathbf{H}_{\mathbf{Cx}}+\mathbf{N}^{\mathbf{T}}\mathbf{H}_{\mathbf{Nx}}\right)\mathbf{x}. \qquad (3)
\end{aligned}
$$

Requirement (2) in the approximation (3) leads to several conditions for the case when $\mathbf{N}$ is a distributed multivariate normal with $\mathrm{E}[\mathbf{N}]=\mathbf{0}$, $\mathrm{Cov}[\mathbf{N}]=\boldsymbol{\Omega}$ in the production or market.

Initially, $\mathbf{B}^{\mathbf{T}}=\mathrm{E}[\mathbf{B}_{\mathbf{x}}^{\mathbf{T}}+\mathbf{x}^{\mathbf{T}}\mathbf{H}_{\mathbf{xx}}+\mathbf{C}^{\mathbf{T}}\mathbf{H}_{\mathbf{Cx}}+\mathbf{N}^{\mathbf{T}}\mathbf{H}_{\mathbf{Nx}}]$ for arbitrary $\mathbf{x}$ is a necessary condition for the linearity of the system which is equivalent to

$$\mathbf{H}_{\mathbf{xx}}=\mathbf{0} \quad \text{and} \quad \mathbf{B}^{\mathbf{T}}=\mathbf{B}_{\mathbf{x}}^{\mathbf{T}}+\mathbf{C}^{\mathbf{T}}\mathbf{H}_{\mathbf{Cx}}. \qquad (4)$$

Introduction of a stability measure or non-centrality parameter for the linearity, referred to as the signal to noise ratio by Taguchi, allows the expression for $\mathbf{B}^{\mathbf{T}}$ to be generally extended as

$$(\mathbf{B}_{\mathbf{x}}^{\mathbf{T}}+\mathbf{C}^{\mathbf{T}}\mathbf{H}_{\mathbf{Cx}})(\mathbf{H}_{\mathbf{Nx}}^{\mathbf{T}}\boldsymbol{\Omega}\mathbf{H}_{\mathbf{Nx}})^{-1}(\mathbf{Bx}+\mathbf{H}_{\mathbf{Cx}}^{\mathbf{T}}\mathbf{C})$$

where $\mathbf{H}_{\mathbf{Nx}}^{\mathbf{T}}\boldsymbol{\Omega}\mathbf{H}_{\mathbf{Nx}}=\mathrm{Cov}[\mathbf{B}^{\mathbf{T}}]$. Then the local stability of linearity can be achieved by satisfying the following condition,

$$\mathbf{H}_{\mathbf{Cx}}^{\mathbf{T}}\mathbf{C}=s\mathbf{u}-\mathbf{Bx} \qquad (5)$$

where $\mathbf{u}$ is the eigenvector corresponding to the maximum eigenvalue of $\mathbf{H}_{\mathbf{Nx}}^{\mathbf{T}}\boldsymbol{\Omega}\mathbf{H}_{\mathbf{Nx}}$ and $s=\|\mathbf{Bx}+\mathbf{H}_{\mathbf{Cx}}^{\mathbf{T}}\mathbf{C}\|$, which can be interpreted as the multivariate generalization of Taguchi's sensitivity since s becomes a scalar or $\mathbf{B}$ in the case $p=1$. This means we can achieve both stability and the target $\mathbf{B}$ for the case $p=1$, and if $p>1$ the direction of the optimal $\mathbf{B}$ from the viewpoint of the stability must be consistent with that of $\mathbf{u}$.

On the other hand, the expectation of the bias term of (3) vanishes if

$$\mathrm{E}\left[Y_0 + \mathbf{B}_\mathbf{C}^\mathbf{T}\mathbf{C} + \mathbf{B}_\mathbf{N}^\mathbf{T}\mathbf{N} + \frac{1}{2}(\mathbf{C}^\mathbf{T}, \mathbf{N}^\mathbf{T})\begin{pmatrix}\mathbf{H}_\mathbf{CC} & \mathbf{H}_\mathbf{NC}^\mathbf{T} \\ \mathbf{H}_\mathbf{NC} & \mathbf{H}_\mathbf{CC}\end{pmatrix}\begin{pmatrix}\mathbf{C} \\ \mathbf{N}\end{pmatrix}\right]$$
$$= Y_0 + \mathbf{B}_\mathbf{C}^\mathbf{T}\mathbf{C} + \frac{1}{2}\operatorname{tr}(\boldsymbol{\Omega}\mathbf{H}_\mathbf{CC})$$
$$= 0$$

or

$$\mathbf{B}_\mathbf{C}^\mathbf{T}\mathbf{C} = -\left\{Y_0 + \frac{1}{2}\operatorname{tr}(\boldsymbol{\Omega}\mathbf{H}_\mathbf{CC})\right\}. \tag{6}$$

Furthermore the variability of the bias term caused by the noise factor $\mathbf{N}$ can be determined by minimizing of the following quadratic form,

$$(\mathbf{B}_\mathbf{N}^\mathbf{T} + \mathbf{C}^\mathbf{T}\mathbf{H}_\mathbf{NC}^\mathbf{T})\boldsymbol{\Omega}(\mathbf{B}_\mathbf{N} + \mathbf{H}_\mathbf{NC}\mathbf{C}) \tag{7}$$

or

$$\mathbf{H}_\mathbf{NC}\mathbf{C} = t\mathbf{v} - \mathbf{B}_\mathbf{N} \tag{8}$$

where $\mathbf{v}$ is the eigenvector corresponding to the minimum eigenvalue of $\boldsymbol{\Omega}$ and $t = \|\mathbf{B}_\mathbf{N} + \mathbf{H}_\mathbf{NC}\mathbf{C}\|$, which is also regarded as a sensitivity. Thus, in the case $p + 1 + n < c$ given the target sensitivities s_0 and t_0, the conditions (5), (6) and (8) can be satisfied by the solution $\mathbf{C_0}$ of the following system of equations,

$$\mathbf{H}_\mathbf{Cx}^\mathbf{T}\mathbf{C_0} = s_0\mathbf{u} - \mathbf{B_x}$$
$$\mathbf{B}_\mathbf{C}^\mathbf{T}\mathbf{C_0} = -\left\{Y_0 + \frac{1}{2}\operatorname{tr}(\boldsymbol{\Omega}\mathbf{H}_\mathbf{CC})\right\}$$
$$\mathbf{H}_\mathbf{NC}\mathbf{C_0} = t_0\mathbf{v} - \mathbf{B}_\mathbf{N}.$$

Therefore a set of solutions can will be formally represented by using the g-inverse matrix, which is not uniquely determined in the case $c > p + n + 1$, as

$$\mathbf{C_0} = \begin{pmatrix}\mathbf{H}_\mathbf{Cx}^\mathbf{T} \\ \mathbf{B}_\mathbf{C}^\mathbf{T} \\ \mathbf{H}_\mathbf{NC}\end{pmatrix}^{-}\begin{pmatrix} s_0\mathbf{u} - \mathbf{B_x} \\ -\left\{Y_0 + \frac{1}{2}\operatorname{tr}(\boldsymbol{\Omega}\mathbf{H}_\mathbf{CC})\right\} \\ t_0\mathbf{v} - \mathbf{B}_\mathbf{N}\end{pmatrix}.$$

Interpretation of the Taguchi method using engineering models or the fusion of Taguchi methods and mathematical modeling is particularly important for computer simulation-based experiments in which fluctuation can be incorporated by using only noise factors to acquire information on the stability or robustness of systems.

5 Future Issues

Although this paper treats only the statistical aspect of the optimization stage of technology development, this method of thinking should also be systematically developed for each stage in technology development in the future and not only real observations but also virtual observations using computer simulations will be essential in future statistical methods, where the random variation itself will need to be designed by statistical engineers after selecting or translating the values of their society.

References

1. Box, G.E.P., Hunter, W.G. and Hunter, J.S. (1978) *Statistics for Experiments: An Introduction to Design, Data Analysis and Model Building*, John Wiley & Sons.
2. Fisher, R.A. (1935) *The Design of Experiments*, Oliver and Boyd.
3. Galton, F. (1883) *Inquiries into Human Faculty and its Development*, Macmillan.
4. Pearson, K. (1892) *The Grammar of Science*, Walter Scott.
5. Shewhart, W. (1931) *Economic Control of Quality of Manufactured Product*, Van Nostrand.
6. Taguchi, G. (1986) *Introduction to Quality Engineering*, Asian Productivity Organization.
7. Wu, C.F.J. and Hamada, M. (2000) *Experiments, Planning, Analysis, and Parameter Design Optimization*, John Wiley & Sons.

Informed Systems Approach

Andrzej P. Wierzbicki

21st Century COE Program: Technology Creation Based on Knowledge Science, JAIST (Japan Advanced Institute of Science and Technology), Japan
and National Institute of Telecommunications, Poland;
email: andrzej@jaist.ac.jp

Summary. Starting with a discussion of the impact of systems science and technology on the current change of understanding the world, this paper describes the need of change of many basic paradigms and theories. This includes the need of a new understanding of systems science. However, this understanding is related to interdisciplinary knowledge integration; thus, it is necessary first to review new *micro-theories* of knowledge creation. After this review, the concept of intercultural and interdisciplinary, informed systemic integration is presented and followed by conclusions.

1 The impact of systems science and technology on current change of understanding the world

In the 20th century, physics contributed first the concepts of *relativism* and *indeterminism*. Then biology contributed the concept of *punctuated evolution*, see, e.g., K. Lorentz (1965), which amounts to an *empirical justification* of *the emergence principle: on subsequent layers of systemic complexity, new phenomena, properties and concepts arise, irreducible to the behavior of the elements of lower layers.* Because of the ideological criticism of the concept of evolution, the empirical justification was not sufficient; however, a *rational substantiation* of this principle was provided additionally soon by *hard systems science*, in particular, by nonlinear dynamic systems analysis and computational engineering that have lead to the *theory of chaos* — both in its deterministic variant, see e.g. J. Gleick (1987) and stochastic variant, see e.g. I. Prigogine (1984) — and to the *theories of complexity*, either *systemic complexity* with order and new phenomena emerging out of chaos because of complexity, or of *computational complexity* assessing the difficulty of large computational tasks.

Thus, we can say that systems science motivated the current change of understanding the world. Technology also contributed to this by giving a *pragmatic justification* of the emergence principle. In particular, telecommunications and computer networks gave first practical examples of very large multilayered systems with new functions emerging at each layer, irreducible to and independent of the functioning of lower layers (in so called ISO-OSI stack of network protocols, see, e.g., Wierzbicki *et al.*, 2003). Additionally, the informational systemic technology, e.g., computerized systems of multiple criteria decision support, has drawn attention to interactive decision making and to unconscious aspects of human behavior, thus contributed to the revolutionary situation in epistemology we observe today and discuss in further text.

We live at a time of civilization change, while the new era has many names: *postindustrial, service, third wave, post-capitalist, information(al) society, knowledge-based economy,* etc. There are diverse ways of periodization of civilization eras; the classical is of third wave type (e.g., agricultural-industrial-informational; antique-modern-postmodern). But a more justified way of periodization is based on following the example of historians, e.g., F. Braudel (1979), who analyzes the era of preindustrial emergence of capitalism, the time of print, banking and geographical discoveries, in Europe 1440–1760, and illustrates this way the concept of a *long duration historical structure.* We can thus discuss the following three modern eras of increasingly globalized civilization:

- Preindustrial civilization: print, banking and geographical discoveries, from Johann Gutenberg: 1440–1760
- Industrial civilization: steam, electricity, transportation mobility, from James Watt: 1760–1980
- Informational and knowledge civilization: computer networks, mobile telephony, knowledge engineering and knowledge economy, from Internet: 1980–2100(?)

For the substantiation of the date 2100 see the book *Creative Space,* A.P. Wierzbicki and Y. Nakamori (2005).

As indicated by F. Braudel, each of such civilization eras is characterized by an unique perspective of perceiving the world, expressed also in the *episteme*, a unique way of constructing knowledge formed during a given civilization era, see M. Foucalt (1970), but earlier in a *conceptual platform* (called also *cultural platform*) of basic concepts characteristic for a new era, preceding even the beginning of a new era and thus also the formation of an episteme, see A.P. Wierzbicki (1988). We already indicated some of such basic concepts for the new era of knowledge and informational civilization: *relativism, indeterminism, punctuated evolution, emergence principle, chaos theory and complexity theory, systemic analysis and integration,* etc. Thus, the conceptual platform is already much advanced, even if the new episteme is not yet formed.

2 The need of change of paradigms and theories

The change of civilization eras motivates also the change of basic paradigms and theories; we give below some examples of recently observed or just needed change:

- In information technology, we do not any longer believe in the validity of reduction to the Turing machine; human mind cannot be modelled as a giant computer, we assume emergence and human-centered approach to information systems.
- In economics, market combined with high technology resulted, as it is well known, in the end of the communist system; but also resulted, as it is less well known, in the end of classical free market theory, which is not any longer applicable to high technology markets (prices on high technology markets exceed marginal production costs hundreds of times, a monopolistic or oligopolistic behavior predominates). Thus, we need a new economic theory.
- In sociology, all postmodernist development forms actually an antithesis to the Comtian postulate of objectivity of sociology, and the postmodern sociology of science believes that all knowledge is subjective — results from a discourse, is constructed, negotiated, relativist, is motivated by the desire for power and money. However, such an episteme is not internally consistent and expresses only a crisis in sociology at the end of industrial civilization era.[1] On the other hand, in the knowledge civilization era we need sociology which would be able to understand that technology creation cannot be successful without striving to objectivity treated not as an attainable absolute, but as an useful value; thus, we need a new sociology, based on a synthesis of subjectivity, intersubjectivity and objectivity.
- In epistemology, most philosophers of 20th century accepted, often subconsciously, the Wittgensteinian prohibition *"wovon man nicht sprechen kann, darüber muss man schweigen"*, meaning *"do not speak about methaphysics"*. However, in the last 15 years at least 7 new *micro-theories of knowledge creation* emerged, all based on the concepts of *tacit knowledge, intuition, emotion and instincts*, thus metaphysics. Therefore, we need also a new epistemology.

We use the concept of a *micro-theory*, because the epistemology of 20th century concentrated either on knowledge validation or on macro-theories of knowledge creation on a grand historical scale, as in the theory of T.S. Kuhn

[1] Postmodern sociology treats science as a social discourse. What happens if we apply this to sociology itself? A paradox: *sociology is a social discourse about itself*, see H. Kozakiewicz (1992). Hard science and technology represent different cultural spheres than sociology. But then, if the postmodern sociology of science (represented, e.g., by B. Latour, 1990) tells hard science and technology that they do not value truth and objectivity, only power and money, then this opinion is self-reflecting on sociology, signifies its internal crisis.

(1970); but this did not lead to conclusions *how to create knowledge for today and tomorrow*, for the needs of knowledge economy. Therefore, most of new micro-theories of knowledge creation came from outside philosophy, from systems science and management science. Since the 20th century was dominated by the Wittgensteinian prohibition not to speak about metaphysical matters, these theories amount to a scientific revolution; we shall return to them later in relation to a new understanding of systems science.

3 New understanding of systems science

Systems engineering is as old as industrial civilization. The invention of James Watt concerned not the steam engine as such (that was known earlier, but had unstable rotary speed and tended to fall apart) but a mechanical control feedback system for stabilizing its rotary speed. The first systematic definitions of the concepts of system and systemic analysis were given by A. Comte (1840):

- *System is an assembly of parts together with interrelations between them*;
- *Systemic analysis means achieving understanding of the system both as a whole and as interconnected parts.*

A. Comte used these concepts to justify his assumptions that social sciences can be made as objective as physics, but this lead to the concept of objective laws of history of K. Marx and subsequent political indoctrination. On other hand, these concepts were taken up and enriched by technology — telecommunications, control engineering and robotics, computer technology and networks. This development contributed to the concepts of:

- *feedback* (1920–40);
- *dynamic behavior of complex systems with sliding motion and deterministic chaotic behavior* (1950–65);
- *interdisciplinary mathematical modelling of systems* (1940–80);
- *hierarchical multi-layered systems* (1965–85)

Actually, neither N. Wiener (1948) nor W.R. Ashby (1958) invented the concept of feedback; they only defined *cybernetics* by accepting and generalizing to living organisms or societies the concepts developed by technology. Parallel, L. Bertalanffy (1956) gave an antithesis to cybernetics by defining *general systems theory,* based on the concepts of open systems and of synergy — that the whole is bigger than the sum of its parts. However, synergy was known earlier to technologists, who treat it as an obvious concept, since they are motivated by the joy of creating technical systems and the complete system means always more to them than the sum of its parts.

Operations research simplified for the needs of management science mathematical modelling used originally in systems engineering; the simplification excluded, between others, the aspects of dynamics and feedback. British management scientists, starting with P. Checkland (1978) and his followers, developed *soft systems approach* by defining hard systems approach as equivalent to operations research (actually, to an approach dominant in management science and operations research at that time, consisting in *finding the efficient means to achieve a given end*). They used this rather narrow definition and the allegedly *functionalist character* of hard technological systems thinking to criticize the hard systems approach; thus, the soft systems approach developed actually as an anti-technological and anti-hard approach. In their critique, they had to attribute to somebody the important concepts of *systems dynamics*, introduced first actually by V. Bush, the inventor of the first (analog) computer (1932), and of *feedback*, applied first as a phenomenon by J. Watt and then introduced as a concept by H. Nyquist (1932) and other telecommunication specialists. Being anti-hard, the British management scientists attributed — see M. Jackson (2000) — the concept of systems dynamics incorrectly to J. Forrester who originally used the concept of industrial dynamics and later (1960) borrowed from technology the concept of systems dynamics. Similarly, they attributed the concept of feedback incorrectly to N. Wiener (1948), who just has drawn the attention to the fact that feedback is a phenomenon known also in biology. We should add that while Forrester in 1960 used the elementary concepts of systems dynamics, such as block diagrams used much earlier by both Bush and Nyquist, the technological, hard systems dynamics had in 1960 already far developed the study of nonlinear systems complexity, leading soon to the theory of deterministic chaos.

Meanwhile, soft systems approach used also the developments of sociology that turned against Comte by denying the role of objectivity and saying that all knowledge is constructed, negotiated, relative, subjective. Although soft systems approach was motivated as anti-hard and anti-technologist, it developed many valuable methods of social debate, of intersubjective systemic analysis, such as interpretive, emancipatory, critical systems thinking. For example, the *Soft Systems Methodology (SSM)* introduced by P. Checkland, see (1978), stresses listing diverse perspectives, so-called *Weltanschauungen*, problem owners, and following an open debate representing these diverse perspectives. Actually, when seen from a different perspective, that of hard mathematical model building, SSM — if limited to its systemic core — is an excellent approach, consistent with the lessons derived earlier from the art of modelling engineering systems.

More doubts arise when we consider the paradigmatic motivation of SSM. SSM is presented by P. Checkland as a general method, applicable in interdisciplinary situations; but a sign of its paradigmatic character is his opinion that soft systems thinking is broader and includes hard system thinking as defined there. But then, should not SSM be also applicable to itself? It includes two *Weltanschauungen*: hard and soft; thus the problem owners of hard *Weltanschauung* should have the right to define their own perspective. However, hard systems practitioners never agreed with the definition of hard systems thinking given by Checkland. He defines hard systems thinking as the belief that *all problems ultimately reduce to the evaluation of the efficiency of alternative means for a designated set of objectives.* On the other hand, hard system technological practitioners say no, they never had such belief; *they are hard because they use hard mathematical modelling and computations*, but for diverse aims, including technology creation, when they often do not know what objectives they will achieve. As a result, hard technological practitioners and soft systems researchers simply do not understand each other.

Thus, *technologists and hard systems thinkers say that the critique by soft systems approach is misinformed*, in particular:

- The correct definition of hard systems approach is *computerized, interdisciplinary analysis of mathematical models of systems* with diverse ends, one of them being simply a better understanding of modelled systems, including understanding of their complexity;
- The critique by soft systems approach concerns operations research as it existed until 1975, does not take into account the developments of hard systems approach after this time nor the achievements of hard dynamic systems analysis even before this time.

The developments of hard systems approach after 1975, unnoticed by soft systems approach, include:

- Interactive decision analysis, assuming the sovereign role and intuitive capabilities of the decision maker, thus avoiding modelling his preferences in detail;
- Soft modelling and computing, including multi-valued logic (fuzzy sets, rough sets, etc.);
- Developments of stochastic systems theory and, in particular, of stochastic theory of chaos (e.g., I. Prigogine et al. 1984).[2]

Thus, a crisis of systems science developed: although it was supposed to be interdisciplinary, it has split into two separate subdisciplines not understanding each other:

- Hard systems scientists disregard valuable soft systems approaches, because they consider the critique by soft systems scientists superficial;
- Soft systems scientists are not properly informed about the possibilities of hard systems approach, because they keep to an incorrect definition of this approach.

[2] While the emergence of chaos in deterministic systems is more difficult to understand — even if it is applied for over 50 years in any computer with a pseudorandom number generator — the emergence of order in stochastic systems is actually rather simple mathematically. Since probabilities sum up to one, a recursive application of a highly nonlinear transformation to a probability distribution increases the highest probability and decreases all others, eventually to zero, while the highest probability converges to one, thus to order.

Japanese researchers tried to synthesize hard and soft systemic approaches, as in *Shinayakana Systems Approach*, (Y. Nakamori and Y. Sawaragi 1990), but this was not fully recognized by soft systems scientists. *A new synthesis, a new understanding of systems science is thus needed.* It should be based on following postulates:

- *It must be open, not excluding any systemic approaches explicitly, by design;*
- *It must be informed, not excluding any systemic approaches tacitly, by bias;*
- *It must be and stay interdisciplinary, not create new disciplinary divisions, e.g. between sociologists and technologists*
- *It must include intercultural understanding, by explicitly discussing diverse disciplinary, national and regional cultures and their contributions to systems science;*
- *Finally, it must be oriented towards systemic integration of interdisciplinary knowledge, the main field of systems science in the era of knowledge civilization.*

Because of the last requirement, we discuss in more detail diverse aspects of new micro-theories of knowledge creation before we return to the new understanding of systems science.

4 New micro-theories of knowledge creation

We mentioned earlier many new *micro-theories of knowledge creation* that can be used to explain how to create knowledge for today and tomorrow, for the needs of knowledge economy. Many of them are directly or indirectly related to Japanese origin (and many to JAIST); they are:

- *Shinayakana Systems Approach* of Y. Nakamori and Y. Sawaragi (1990);
- *SECI Spiral* of I. Nonaka and H. Takeuchi: *The Knowledge Creating Company* (1995)
- *Rational Theory of Intuition* by A.P. Wierzbicki (1997)
- *Process of Regress* by A. Motycka (1998), using the concept of *collective unconscious* of C.G. Jung (1953)
- I^5 *or Pentagram System* by Y. Nakamori (2000)
- *The Management of Distributed Organizational Knowledge* by S. Gasson (2004)
- Several others.

Because their integration is based on the rational theory of intuition, we start with a description of this theory.

4.1 The Rational Theory of Intuition

Intuition has been long regarded — by Plato, Descartes, Kant — as an infallible source of inner truth concerning *a priori synthetic judgments*; but it turned out later (with non-Euclidean geometry and with relativity theory) that such judgments are not necessarily true, intuition is fallible. Nevertheless, in the time of knowledge-based economy we need a rational theory of intuition, because it is a powerful source of innovative knowledge creation, even if it might be fallible and thus must be checked empirically.

This theory uses facts known from modern telecommunications and computational complexity theory about relative complexity of processing vision (pictures, television) and speech (audio signals): the ratio of bandwidth 100 : 1, the ratio of processing complexity at least 10 000 : 1, a picture is worth (at least) ten thousand words.[3] Then we use the following thought experiment: Imagine the period in human evolution when we discovered speech. How did we process signals from our environment just before this discovery? *The discovery of speech was a tremendous evolutionary shortcut: we could process signals* 10^4 *times faster.* This made possible the intergenerational transfer of knowledge and thus started entire evolution of human civilization, but also *slowed down the biological evolution of our brains.* Thus, in evolutionary epistemology *we should carefully distinguish between biological and cultural evolution.*

The discovery of speech was a simplification that accelerated intellectual evolution of humans. As with each simplification, it had also drawbacks. E.g., binary logic was practically discovered as a tool of ideology: "This must be true or false, there is no third way!" But what happened to our old abilities of pre-verbal processing of information? The discovery of speech suppressed these abilities, pushed them down to subconscious or quasi-conscious behavior. Our consciousness, especially its logical and analytical part, became strictly related to speech and verbal articulation. However, we still possess our old preverbal abilities of information processing that are at least 10^4 times stronger than verbal; for the lack of a better word, we call them intuition.

Definition: *Intuition is quasi-conscious and unconscious, pre-verbal and holistic (parallel, distributed) information processing, utilizing aggregated experience, training and imagination and performed by a specialized part of human mind.*

This is a *rational evolutionary definition of intuition* that allows falsifiable (testable experimentally) practical conclusions:

---

[3] This is a lower bound estimate, because human hearing seldom exceeds 20 kHz, while television with 2 Mhz bandwidth is much less precise than human vision; moreover, a quadratic increase of complexity of processing large sets of data is one of the mildest nonlinear increases, computational complexity theory shows that for most difficult problems this increase should be no-polynomial, e.g. exponential.

- For example, a conclusion is that in order to stimulate intuition we must *Suppress Consciousness* (as in Zen meditation or in athletic concentration);
- Another conclusion is the *Alarm Clock Method*: in order to easier generate new ideas, put your alarm clock 10 minutes earlier and think hard upon awakening whether during sleep your mind did not find a solution to your most difficult problem;
- Still another conclusion is *Limit TV*: in order to be more creative, do not saturate your imagination by spending too much time before TV.

Entire Rational Theory of Intuition was developed, with appropriate concepts, definitions, structure of intuitive decision processes, relations to cognitive sciences (right brain–left brain), etc. An essential distinction is between:

- repetitive, *operational* intuitive decisions
- unique (non-repetitive) *strategic* intuitive decision processes, including creative decisions.

Two most important theoretical conclusions from all these considerations are:

- There are two parts of *tacit knowledge*:
  - *intuition*, our intuitive ability of utilising life-long experience and imagination;
  - *emotions*, which actually include also part of explicit knowledge;
- An essential element of intuitive creative processes is the phenomenon of *enlightenment* — having an idea, called also variously *aha, illumination, eureka.*
- *Intuition is powerful but fallible*, we must check our intuitive conclusions in various ways, including empirical falsification attempts.

An application of the rational theory of intuition leads also to the concept of *Creative Space.*

4.2 The concept of *Creative Space*

This concept represents a way of integrating known theories of knowledge creation, based on following assumptions:

- Use at least *three-valued logic* (three levels of ontological elements);
- Use the *epistemological* and the *social* (originally called not quite precisely *ontological*) dimensions of *SECI Spiral* as basic, but add also other dimensions, as in I^5 *System*;
- Carefully define the *nodes* (ontological elements) of *Creative Space* and study possible *transitions* (originally called, also not quite precisely *knowledge conversions*) between the nodes;

- Thus, create a *network-like general model* of creative processes.

We start with the *SECI Spiral* of I. Nonaka and H. Takeuchi (1995) and recall that it consists of four transitions: *Socialization* from individual tacit to group tacit knowledge; *Externalization* from group tacit to group explicit knowledge; *Combination* from group explicit to individual explicit knowledge; and *Externalization* from individual explicit to individual tacit knowledge (for example, the last one occurs during learning by doing), see Fig. 1. However, if we change the binary logic of *tacit or explicit* into trinary: *emotions-intuition-rationality* and extend correspondingly the binary distinction: *individual-group* into trinary: *individual-group-humanity*, then we obtain a generalization of the *SECI Spiral* to the *Creative Space* as shown in Fig. 2.

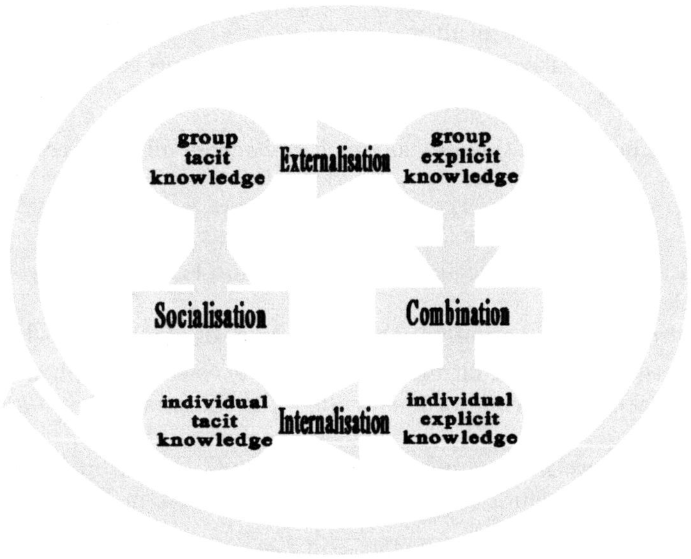

Fig. 1. The *SECI Spiral* (Nonaka and Takeuchi)

The ontology of Creative Space includes its nodes and transitions; all are discussed in more detail in the book *Creative Space*, A.P. Wierzbicki and Y. Nakamori (2005). We give here only a shortened examples of such discussion. For example: the *emotive heritage of humanity* contains not only tacit knowledge elements, also explicit knowledge: music, movies, other art forms — which all influence our creativity. However, their influence is indirect, through emotions, even if we know them explicitly. You might have seen a movie thus you know it explicitly; but its impact on your creative behavior is indirect, emotional. Thus the distinction *emotive-intuitive-rational* is more precise, than *tacit-explicit*. The emotive heritage of humanity contains also an important tacit element — the *collective unconscious* of C.G. Jung (1953).

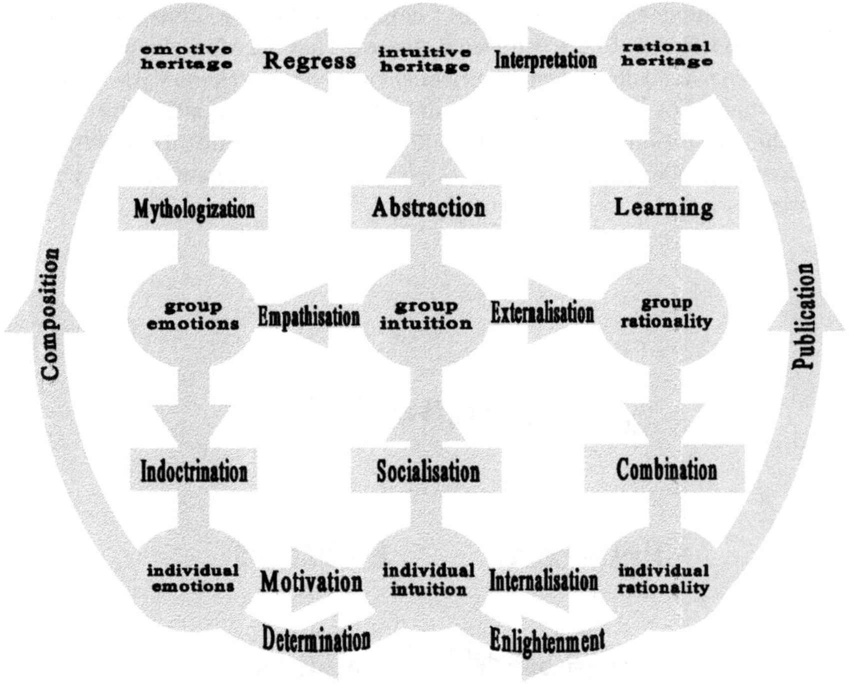

Fig. 2. Basic dimensions of *Creative Space*

Another example: the *intuitive heritage of humanity* is actually equivalent to the Kantian *a priori synthetic judgements*. First Plato, later Descartes and finally Kant have shown that there are true ideas that we have in our mind and they appear obvious to us (ideas of space, time, logic). However, the concepts of non-Euclidean space, relative time, multivalued logic have shown that these ideas are not obvious and not necessarily true. The Rational Theory of Intuition suggests that they are our intuitive heritage of humanity — taught by playing with blocks and by the paradigm of teaching mathematics, extremely valuable for civilization evolution, but mesocosmic (resulting from our perception of the middle scale, we do not directly perceive the micro- or macro-scale of the universe), hence not necessarily true.

4.3 New spirals for creative processes

While the *SECI Spiral* is in some sense preserved in the lower right hand corner of the *Creative Space* in Fig. 2, other nodes and transitions can result in other creative spirals. In fact, the upper left hand corner in Fig. 2 corresponds to the *Regress Theory* of A. Motycka (1998). In a time of crisis of a scientific discipline (e.g. physics before quantum theory) a group of scientists comes intuitively to conclusion that they must find new concepts. They seek such concepts in mathematical intuition, but cannot find them. Thus, they turn to collective unconscious of C.S. Jung in a process that A. Motycka calls *regress.* Archetypes contained in collective unconscious provide them with sources of new ideas which they discuss on emotional level until by empathy they become a part of their intuition. If this is sufficient, mathematical intuition helps them to formalize their ideas. If not, the process of regress is repeated. This can be represented as a spiral, repeated in Fig. 3.

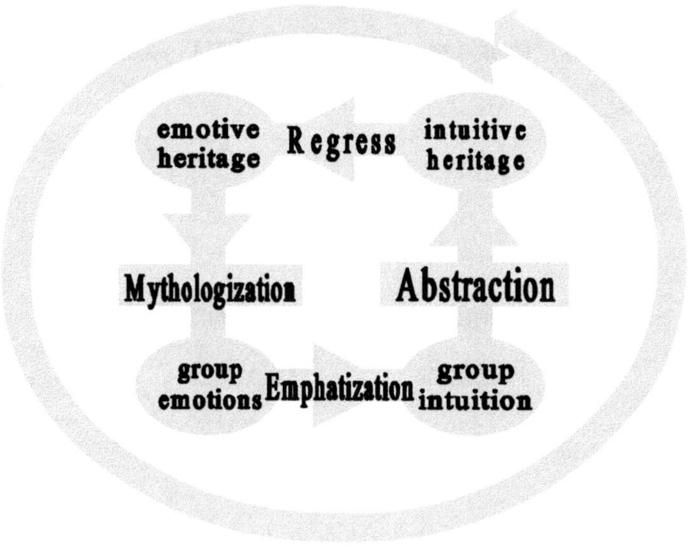

Fig. 3. The *ARME Spiral* (Motycka)

Thus the historical phenomenon of creating new scientific theories in times of a scientific revolution, described by A. Motycka independently from I. Nonaka and H. Takeuchi *SECI Spiral*, can be represented as another spiral in the Creative Space.

T.S. Kuhn argued that scientific revolutions are divided by long phases of normal science development. Neither *ARME Spiral* nor *SECI Spiral*, for all their revolutionary and organizational value, do not describe the process of *normal science or technology creation in academia* (at universities, in research institutes). However, we can also find a spiral in the *Creative Space* — going through the same nodes as the *SECI Spiral*, but through different transitions and in the opposite direction — that describes at least one of the aspects of the known process of normal science and technology creation, see Fig. 4.

The *EDIS Spiral* describes actually a very basic creative process. An individual researcher, after having an intuitive idea in the transition *Enlightenment*, wants to test this idea intersubjectively — though a discussion with his colleagues, in the transition *Debate*. But what we do with the results of such

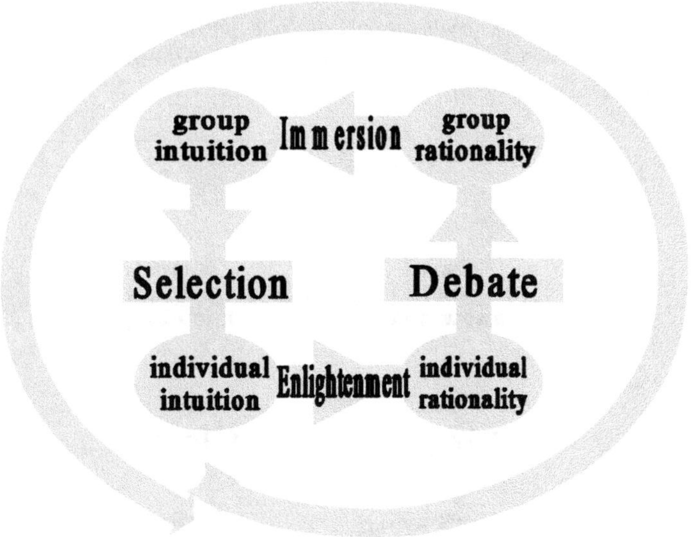

Fig. 4. The *EDIS Spiral* (Wierzbicki)

a discussion? All participants of the *Debate* participate in an *Immersion* of the rational results of discussion into their intuition — from which it follows, actually, that best results would be achieved if the *Debate* would be repeated after a not too long time. The individual researcher might use the comments of his colleagues or not, he makes an intuitive *Selection* of the debated ideas, which might lead to a new *Enlightenment*. The *EDIS Spiral* does not completely describe normal processes of science and technology creation, because equally important are processes involving the analysis of the rational heritage of humanity — simply searching the existing literature of a research object and reflecting on it — and experimental verification of ideas; but we shall show the corresponding spirals later.

However, practically through the same nodes (only differently called by S. Gasson, see 2004) and in the same direction, but with slightly different transitions goes another *organizational* spiral of knowledge creation in market-oriented organizations or companies, the *OPEC Spiral*, see Fig. 5. This spiral was suggested by S. Gasson as an occidental variant responding to the more oriental variant of the *SECI Spiral* and describing the purposeful knowledge creation processes in Western organizations: first set *Objectives*, the develop *Process*, go through *Expansion* (actually, very similar to *Enlightenment*) and summarize through *Closure*; if the organization does not have enough knowledge distributed between its members, it will hire external experts. While very convincing and valuable, the *OPEC Spiral* (and, actually, also the *SECI Spiral*) does not note that every knowledge creation process uses some part of humanity intellectual heritage. This gives rise to grave questions: *What is the economic value of the intellectual heritage of humanity? Would not an unlimited privatization of knowledge result in a pollution of intellectual heritage of humanity — similar to the pollution of natural environment resulting from its unlimited privatization?*

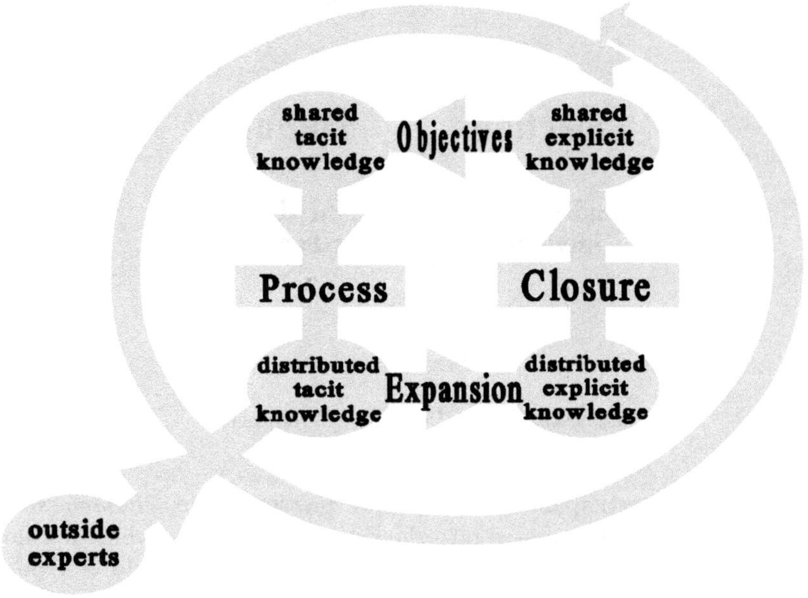

Fig. 5. The *OPEC Spiral* (Gasson)

In Fig. 6 we recall the I^5 *System (Knowledge Pentagram)* of Y. Nakamori (2000). Without describing it in detail it deserves, we note only that the nodes of the pentagram might be interpreted as additional dimensions of the *Creative Space*: while the nodes *Intelligence* and *Involvement* might be interpreted as corresponding to the two basic dimensions *epistemological* and *social* of the *Creative Space*, the nodes *Imagination, Intervention, (Systemic) Integration* indicate further dimensions. Upon further analysis, it is possible — see A.P. Wierzbicki and Y. Nakamori, *op. cit.* — to identify yet another 5 dimensions: *Abstraction, Objectivity, Hermeneutic Reflection, Cross-Cultural and Organizational.* From the total 10 dimensions, we comment here only on selected ones.

The *Objectivity* dimension involves at least three levels: *Subjectivity, Intersubjectivity* and *(Informed) Objectivity.* The last one cannot obviously be interpreted as the full correspondence to empirical facts postulated by positivism, since every measurement is distorted by the act of measuring; informed objectivity means accepting this limitation, but nevertheless testing our fallible, intuitive ideas by experiments. The normal process of knowledge creation using experiments is represented in Fig. 7: upon generating an intuitive idea

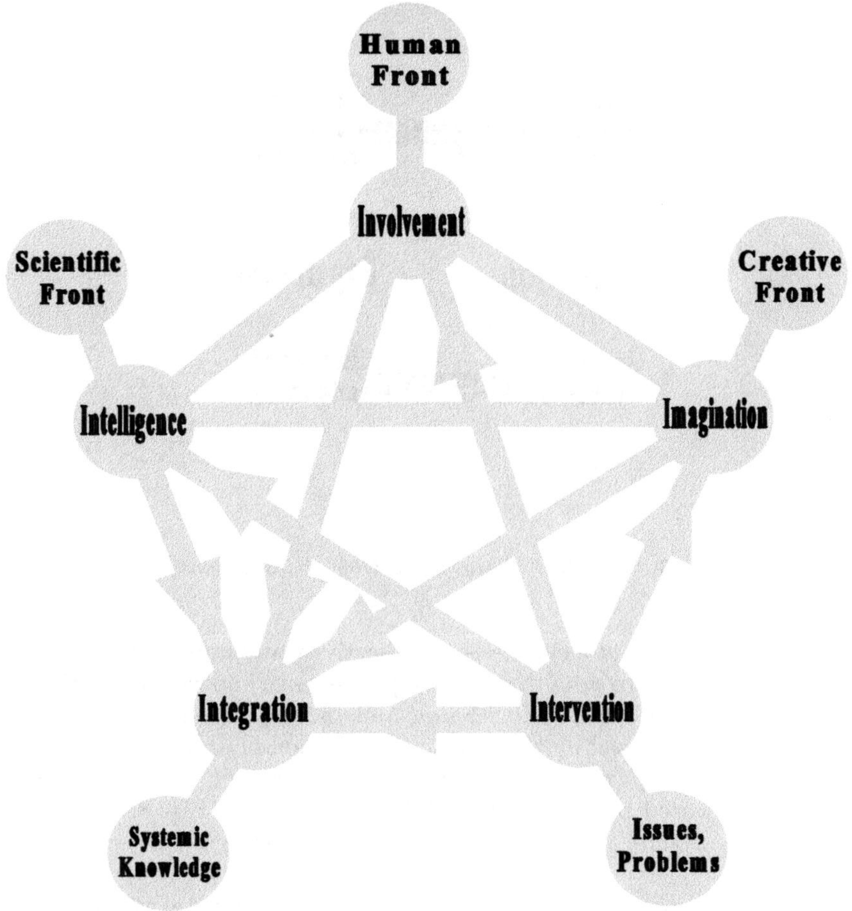

Fig. 6. The I^5 *System (Knowledge Pentagram)* of Nakamori

in the transition *Enlightenment*, an experimental researcher tests this idea by a suitably designed experiment. However, a rational record of experimental results does not suffice, as it is known by any experimental researcher; we must compare these results with our experimental intuition in a transition called *Interpretation*; then we can perform the *Selection* of conclusions that help mostly in generating new ideas.

Technology cannot be created without informed objectivity. We, technologists, understand that absolute truth and absolute empirical accuracy do not exist, but we try to be as empirically accurate, as objective as possible; otherwise we would not be successful in creating technology, and we, technologists, are actually motivated by the creative joy of techne (old Greek word for crafts and arts). Sociology of science condemns objectivity, but we need sociology for a better understanding of coming knowledge civilization era; however, can

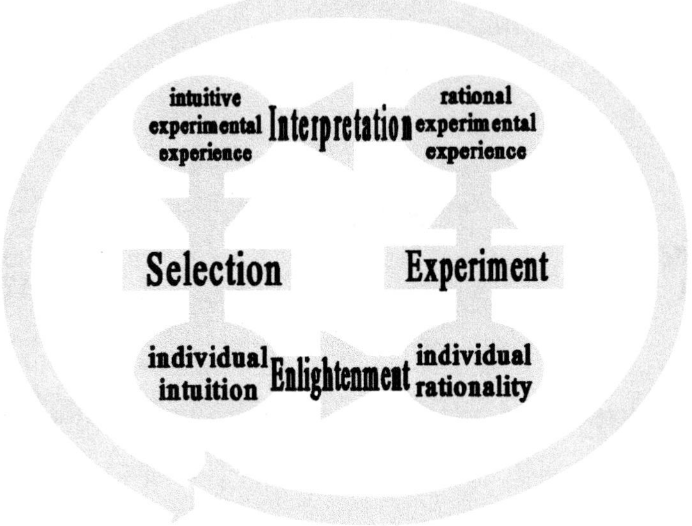

Fig. 7. The *EEIS Spiral* using *Objectivity* dimension

sociology help us without a synthesis of subjectivity and informed objectivity? Hence it is important to note that normal processes of knowledge creation can combine both the debating *EDIS Spiral* and the experimental *EEIS Spiral*, since they are both based on the intuitive generation of an idea in the transition *Enlightenment*; we can switch between these spirals or even perform them parallel, as indicated by the *Double EDISEEIS Spiral* in Fig. 8. The *Double EDISEEIS Spiral* indicates a way of integrating at least intersubjectivity with objectivity.

However, the intersubjective debate and the experimental verification do not exhaust main ways of normal knowledge creation. There is a third, even more basic way: the individual study of literature related to a given object of study and the reflection about this material. Humanities created a special field of study of such reflection, called *hermeneutics*, see, e.g., H.G. Gadamer (1975); however, this basic way of normal knowledge creation is not restricted to humanities, is in fact used also by hard science and technology. Thus, a different way of *closing the hermeneutic circle* is represented in Fig. 9. After having an idea in the transition *Enlightenment*, an individual researcher searches the rational heritage of humanity — the books, the journals, the web — in the transition *Analysis* and thus obtains a rational perception of his object of study. But this does not suffice, he must *immerse* these results into his individual intuitive perception of the object of study in a transition that is essentially hermeneutic, hence called *Hermeneutic Immersion*. Finally, a purely intuitive transition *Reflection* gives basis for the creation of new ideas.

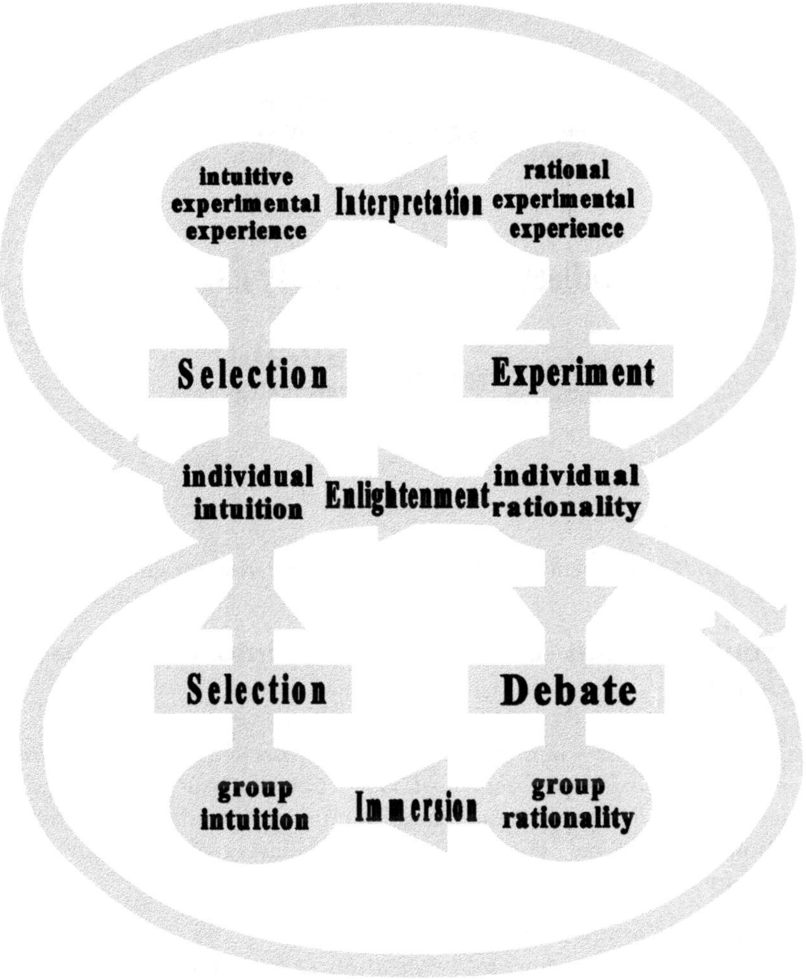

Fig. 8. The *Double EDISEEIS Spiral* of intersubjective and objective knowledge creation and justification

When reflecting upon our experience in normal science creation, we can conclude that these three spirals — the hermeneutic *EAIR Spiral*, the experimental *EEIS Spiral*, the intersubjective *EDIS Spiral* — represent the most frequent ways of normal, academic creativity. They can be switched or performed in parallel; in order to stress this fact, we can represent them also as a triple spiral, or a *Triple Helix* of normal science and technology creation, see the book *Creative Space* of A.P. Wierzbicki and Y. Nakamori, *op. cit.* In this book, also other spirals and processes of science and technology creation are discussed.

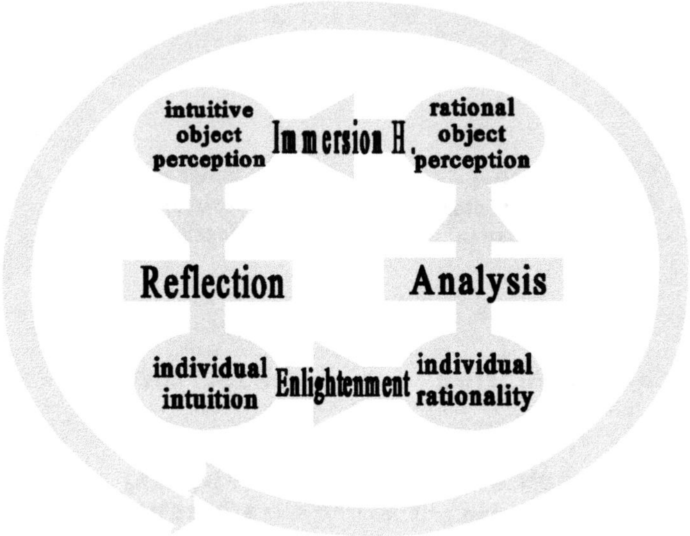

Fig. 9. The hermeneutic *EAIR Spiral* using *Reflection* dimension

5 Intercultural and interdisciplinary, informed systemic integration

We stressed before that a new synthesis of systems science should be *intercultural, interdisciplinary, open, informed.* But what is synthesis or integration? It is a new perception of the whole, based upon good knowledge of details and on intuition and wisdom. The theory of knowledge creation outlined above helps us to understand how to achieve integration:

- through *intersubjective debate (EDIS Spiral);*
- through *objective experimental verification (EEIS Spiral);*
- through *hermeneutic reflection (EAIR Spiral),*

in all cases with a deliberate use of our intuitive abilities. How to support such integration?

We might use the concept of *Creative Environments*: intelligent computer systems to support either selected transitions in Creative Space, or phases of creative processes, or entire creative processes of specific character. The concept of *Creative Environment* is similar to the concept of *Ba* by I. Nonaka, see Von Krogh *et al.* (2000), but stresses more computerized intelligence used to support creativity.

Some computer tools to support creativity already exists, such as word processors that support the transitions *Publication* or *Composition.* Since there exists also a large experience in constructing computerized *decision support systems*, see, e.g., Wierzbicki *et al.* (2000), it can be used for *creativity support systems*. Consider, therefore, the possibility of supporting informed systemic integration:

- A Creative Environment to support good, deep, immersed (Double) intersubjective debate;
- A Creative Environment to support objective experimental verification;
- A Creative Environment to support gathering research materials and hermeneutic reflection.

This is all related to a new definition of systems science proposed here:

Definition: *Future systems science will be the discipline concerned with methods of intercultural and interdisciplinary integration of knowledge, including soft intersubjective and hard objective approaches, open and, above all, informed.*

Intercultural means here explicit account for national, regional, disciplinary cultures, trying to overcome the incommensurability of cultural perspectives. Interdisciplinary approach has been a defining feature of systemic analysis since Comte, but has been gradually lost in the division between soft and hard approaches. Open means pluralist, as stressed by soft systems approach, not excluding by design any cultural or disciplinary perspectives. Informed, as stressed by hard systems approach, means not excluding any such perspectives by ignorance or by disciplinary paradigmatic belief.

6 Conclusions

In the beginning stages of new era of knowledge civilization, it is necessary to change many paradigms and theories, including new approaches to economics, to sociology, to epistemology. Especially important is a new approach to systems science, based on the postulates of intercultural, interdisciplinary, open and *informed* systemic integration. New *micro-theories* of knowledge creation processes can help in achieving such systemic synthesis.

References

1. Ashby, W.R. (1958) Requisite variety and its implications for the control of complex systems. *Cybernetica* **1**:83–99.
2. Bertallanfy, L. (1956) General systems theory. *General Systems* **1**:1–10.
3. Braudel, F. (1979) *Civilisation matérielle, économie et capitalisme, XV-XVIII siécle.* Armand Colin, Paris.
4. Checkland, P.B. (1978) The origins and nature of "hard" systems thinking. *Journal of Applied Systems Analysis* **5**:99.
5. Comte, A. (1830–42) *La mèthode positive en seize leçons* (Polish translation: Metoda pozytywna w 16 wykładach, PWN 1961, Cracow).

6. Findeisen, W., Bailey, F.N., Brdyś, M., Malinowski, K., Tatjewski, P., Woźniak, A. (1980) *Control and Coordination in Hierarchical Systems.* J. Wiley and Sons, Chichester.
7. Foucalt, M. (1972) *The Order of Things: an Archeology of Human Sciences.* Routledge, New York.
8. Gadamer, H.-G. (1975) *Warheit und Methode. Grundzge einer philosophishen Hermeneutik.* J.B.C. Mohr (Siebeck), Tbingen.
9. Gasson, S. (2004) The management of distributed organizational knowledge. In Sprague, R.J. (ed) *Proceedings of the 37 Hawaii International Conference on Systems Sciences (HICSS 37).* IEEE Computer Society Press, Los Alamitos, Ca.
10. Gleick, J. (1987) *Chaos: Making a New Science.* Viking Penguin, New York.
11. Jung, C.G. (1953) *Collected Works.* Pantheon Books, New York.
12. Kozakiewicz, H. (1992) Epistemologia tradycyjna a problemy współczesności. Punkt widzenia socjologa (in Polish, Traditional epistemology and problems of contemporary times. Sociological point of view). In Niżnik, J, ed. *Pogranicza epistemologii* (in Polish, *The Boundaries of Epistemology*). Wydawnictwo IFiS PAN, Warsaw.
13. Kuhn, T.S. (1970) *The Structure of Scientific Revolutions.* Chicago University Press, Chicago (2nd ed.).
14. Latour, B. (1990) Science in action, in: Postmodern? No, simply a-modern! Steps towards an anthropology of science. *Studies in the History and Philosophy of Science* **21**.
15. Lorentz, K. (1965) *Evolution and Modification of Behavior: a Critical Examination of the Concepts of the "Learned" and the "Innate" Elements of Behavior.* The University of Chicago Press, Chicago.
16. Motycka, A. (1998) *Nauka a nieświadomość* (Science and Unconscious, in Polish). Leopoldinum, Wrocław.
17. Nakamori, Y. and Sawaragi, Y. (1990) Shinayakana systems approach in environmental management. *Proceedings of 11th World Congress of International Federation of Automatic Control,* Tallin. Pergamon Press, vol. 5 pp. 511–516.
18. Nakamori, Y. (2000) Knowledge management system toward sustainable society. *Proceedings of First International Symposium on Knowledge and System Sciences,* JAIST, pp. 57–64.
19. Nonaka, I. and Takeuchi, H. (1995) *The Knowledge Creating Company. How Japanese Companies Create the Dynamics of Innovation.* Oxford University Press, New York.
20. Nyquist, H. (1932) Regeneration theory. *Bell System Technical Journal* **11**:126–147.
21. Prigogine, I. and Stengers, I. (1984) *Order Out of Chaos.* Bantam, New York.
22. Von Krogh, G., Ichijo, K. and Nonaka, I. (2000) *Enabling Knowledge Creation.* Oxford University Press, Oxford.
23. Wiener, N. (1948) *Cybernetics or Control and Communication in the Animal and the Machine.* MIT Press, Cambridge, Mass.
24. Wierzbicki, A.P. (1988) Education for a new cultural era of informed reason. In Richardson, J.G. (ed) *Windows of Creativity and Inventions,* Lomond, Mt. Airy, PA.
25. Wierzbicki, A.P. (1997) On the role of intuition in decision making and some ways of multicriteria aid of intuition. *Multiple Criteria Decision Making* **6**:65–78.

26. Wierzbicki, A.P., Makowski, M. and Wessels, J. (2000) *Model-Based Decision Support Methodology with Environmental Applications.* Kluwer Academic Publishers, Boston-Dordrecht.
27. Wierzbicki, A.P., Jankowski, P., Łoniewski, D. and Wójcik, G. (2003) *Computer Networks — a Multimedia Textbook (in Polish), Sieci komputerowe: podręcznik multimedialny.* Instytut Łączności, Ośrodek Informatyki and Politechnika Warszawska, Ośrodek Kształcenia na Odległość, Witryna OKNO, Warsaw.
28. Wierzbicki, A.P. and Nakamori, Y. (2005) *Creative Space.* Springer Verlag, Berlin.
29. Wittgenstein, L. (1922) *Tractatus logico-philosophicus.* Cambridge, UK.

Combinatorial Usage of QFDE and LCA for Environmentally Conscious Design

— Implications from a case study —

Tomohiko Sakao[1], Kazuhiko Kaneko[2], Keijiro Masui[3], and Hiroe Tsubaki[4]

[1] Darmstadt University of Technology, sakao@pmd.tu-darmstadt.de
[2] Ebara Corporation, kaneko.kazuhiko@ebara.com
[3] National Institute of Advanced Industrial Science and Technology, k-masui@aist.go.jp
[4] University of Tsukuba, tsubaki@gssm.otsuka.tsukuba.ac.jp

Summary. Environmentally conscious design (ecodesign) is at present among the key issues for manufacturers. Although many tools for ecodesign exist, how to employ different tools in combination is unavailable. To reveal a way of combinatorial utilization of QFDE (Quality Function Deployment for Environment) and LCA (Life Cycle Assessment), which are characteristic tools for ecodesign, this paper aims at analyzing how designers obtain and utilize different types of outputs from those two tools through a case study of QFDE and LCA on an identical product, an industrial pump. Outputs from QFDE were proved to include their functional roles such as that of an impeller, which supports designers effectively but cannot be obtained from LCA. In contrast, the results produced by LCA are indispensable for obtaining quantitative and objective information on the environmental aspect. Hence, a combinatorial usage of QFDE and LCA is suggested for effective ecodesign.

Key words: combinatorial usage, ecodesign, LCA, QFDE

1 Introduction

During these decades, growing knowledge about relation between products and environmental problems has resulted in an increasing interest on the environment in manufacturing companies. Due to this, researchers have developed a number of ecodesign methodologies or tools for engineering designers. ISO/TR 14062 [1], one of the most known documents for standardization in the ecodesign field, lists about thirty tools applicable to ecodesign, and some of them are actually utilized in companies. However, how to utilize different ecodesign tools in combination with each other has not been discussed enough so far.

* Contributed paper: received date: 10-Jul-05, accepted: 15-Sep-05

There exist two characteristic tools for supporting ecodesign; one is LCA (Life Cycle Assessment) [2], and the other is QFDE (Quality Function Deployment for Environment) developed by the authors [3]. QFDE is a modification from the environmental viewpoint and also an extension of QFD (Quality Function Deployment) [4]. Although some methodological improvements are still being done [5], QFDE has been verified through several case studies [6, 7]. While LCA is a quantitative tool applicable at the later stage of product design, QFDE is a semi-quantitative design tool applicable at the earlier stage.

It should be noted that the two ecodesign tools focused in this paper are related with statistical techniques. As LCA is environmental assessment through a product life cycle, it requires by definition the data in the whole life cycle. It is natural that all the data from so broad areas are not necessarily available, when a kind of estimation must be performed. Furthermore, uncertainty analysis must be needed in some cases. Since QFDE is based on QFD, achieving QFDE also requires statistical techniques to quantify a matrix of the quality table. Furthermore, a kind of simulation based on statistical techniques is carried out to evaluate redesign options in later phases of QFDE. Thus, statistical thinking and usage of statistical methods play an important role in carrying out LCA and QFDE.

In order to pursue a method of utilizing QFDE and LCA in combination, this paper aims at analyzing how designers obtain different types of outputs from these two tools through a case study applying QFDE and LCA to an identical product, an industrial pump.

In Chapter 2, the research motivation will be explained. Following this, Chapter 3 describes the results of the case study where QFDE and LCA are applied to an industrial pump. Chapter 4 discusses implications from those results, and Chapter 5 concludes the paper.

2 The research motivation

2.1 What is ecodesign?

Over a couple of decades, environmental problems have been quite serious. As for the global warming problem, which is one of the most seriously recognized ones, the annual average temperature on the earth has risen considerably [8]. On the waste processing problem, it is predicted that the volume of municipal wastes in OECD countries in the year 2020 will be more than double of that in 1980 [9]. It should be noted that the increase ratio of the municipal waste is much higher than that of the population. Thus, it is recognized that the system of our society at present is not environmentally sustainable. In response to these problems, in 1997, WBCSD (World Business Council on Sustainable Development), an international industrial organization addressed the need of the industries for sustainable development, and presented some strategies with actual activities [10]. At present, the environmental issue has become one of the critical ones for manufacturers: First, they must comply with laws and regulations on, for instance, recycling home appliances, automobiles, and personal computers in Japan [11], and hazardous substances and energy use of products in Europe [12, 13, 14]. Second, customers in some cases demand an environmental performance of a product. Third, they recently have to be responsible for the environment as one factor of CSR (Corporate Social Responsibility). As a result, for product development, the environmental aspect at this moment is regarded as one of the qualities to be assured.

Ecodesign (environmentally conscious design) means incorporating environmental issues such as the global warming, the resource consumption, and the hazardous substances usage into product development. Ecodesign is so important, because it is the activity at design stage that determines most of the environmental impacts through the product life cycle. In ecodesign, "life cycle thinking", which is an idea to see a product from the viewpoint of its whole life from material acquisition to product disposal, is one of the crucial factors, because it allows us to grasp the total environmental impact generated by a product. Thus, it is often required for manufacturers to keep EPR (Extended Producer's Responsibility), which was proposed by OECD in 1994. EPR is an idea in which manufactures should have a certain extent of responsibility even outside the areas of their actual activities such as the disposal stage for home appliance manufacturers.

2.2 Tools for ecodesign

To carry out ecodesign efficiently, product designers need methods and tools. There are actually a number of ecodesign tools, many of them have been utilized in industry. ISO/TR 14062 [1] provides a list for those tools as shown in Table 1. It is favourable for designers that quite a few tools like these are available. However, it must be emphasized that how to utilize those ecodesign

Table 1. The list of ecodesign tools in ISO/TR14062 [1]

1. *Planning stage* Checklists, SWOT analysis Benchmarking QFD (Quality Function Deployment) FMEA (Failure Mode and Effects Analysis) LCA (Life Cycle Assessment), LCC (Life Cycle Costing) Hazard and risk assessment Stakeholder benefits and feasibility analysis	2. *Conceptual design stage* Guidelines and checklists, Manuals Material databases 3. *Detailed design stage* Software and modelling tools Material databases DfA (Design for Assembly), DfDA(Design for Disassembly) Production and process optimization tools Substance lists

tools in combination of each other is not provided so far. In other words, there is a need for "grammar" in ecodesign.

A way of utilizing a set of QFDE (Quality Function Deployment for Environment) and LCA (Life Cycle Assessment) from the list is proposed [15]. Adopting the generic design process by Pahl and Beitz [16], they argue QFDE should be applied to the first step, "Plan and clarify the task", and LCA to the last step, "Prepare production and operating documents" (See Figure 1). It is reasonable because the first step is for translating several types of information, for instance, requirements from the market to product attributes, where QFD is mainly used. Because the last step includes communication and information management, it is appropriate to employ LCA, whose main purposes include communication to other stakeholders. However, how to utilize different ecodesign tools in a combining way is missing, although it is crucial for designers to conduct ecodesign effectively and efficiently.

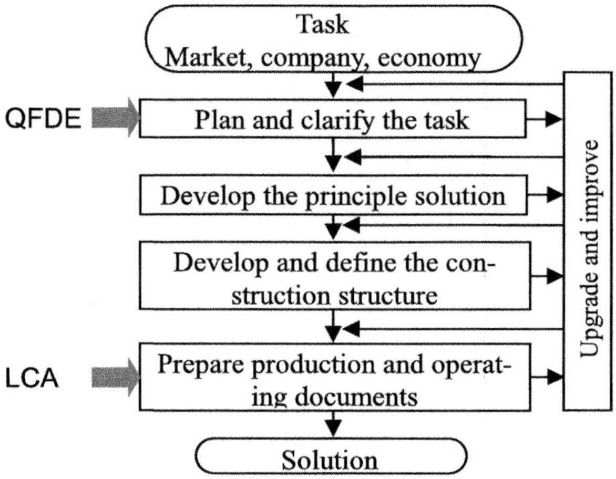

Fig. 1. The applied stages of QFDE and LCA in the design process by Pahl and Beitz [15]

3 Presented case study

3.1 The example product

The example product in this paper is an industrial pump. The product's roles include raising or spraying water for a boiler, fire extinction, or an industrial usage. Figure 2 shows the outlook of the pump used in the case study. It is intended for low lift and large flux. It employs a ball bearing, thus needs no oil supply, which results in easy maintenance. Furthermore, its structure is so simple that the lifetime is relatively long.

Fig. 2. The outlook of the example product

3.2 QFDE of the pump

3.2.1 The QFDE method

QFDE [3] is a methodology to support ecodesign developed by incorporating environmental aspects into QFD. QFD is a methodology used in product planning stage to analyze functions required for a product or the product structure to realize these functions.

QFDE consists of four phases. In Phase I, voices of customers (VOC) with voices of the environment (VOE), and engineering metrics (EM) for traditional and environmental qualities are correlated, while in Phase II EM and part characteristics (PC) are also correlated. Part characteristics can be regarded as function units or components. For both correlations, semi-quantitative scores are used. This information is generated by a group of product developers, which is called a QFDE team in this paper. The outputs of Phase I and II are identification of the function units that should be focused in product design when environmental as well as traditional qualities are considered.

After identifying the important part characteristics, the QFDE team will examine design improvements for the product in Phase III and IV. The team determines redesign target and those changes are expressed by a combination of an EM and a PC to be improved. The team can also evaluate the effects of the design changes on the VOC and VOE using semi-quantitative information represented in the two correlation matrices in Phase I and II.

3.2.2 Phase I

Table 2 shows the results from Phase I of QFDE of the pump. The results show that the "material", "emission of CO_2", and "recycle ratio" are the most important EM (engineering metrics) in its order. Note that the shaded part is environmental. In addition, the scores were selected from 0, 1, 3, 6, and 9 depending their strength, since this was reasonable for the QFDE team.

Table 2. The table for Phase I of QFDE (to be cont. on the next page)

QFD for Environment Phase I / **Customer Requirements**	**Coustomer Weights**	capacity (m^3/min)	total head (m)	rotating speed (rpm)	efficiency (%)	specific speed (—)	suction pressure (MPa)	discharge pressure (MPa)	NPSH (Net Positive Suction Head) (m)
water supply	9	9	9	6	9	1	6	6	6
supplying with high viscosity liquid	1	9	9	6	9	1	6	6	6
supply liquid including solid	1	9	9	6	6	1	6	6	6
supply liquid with fixed quantity	1	9	9		6	1	6	6	6
prevension of cavitation	3					1	9		9
pumping out	1	9	9		6				
utilization as strage pump	9	9	9	6	9		6	6	6
circulation of refrigerant	9	9	3	6	9				
utilization as power pump	1				6			9	
utilization under 50 or 60 Hz area	3				6				
prevent rusting	9								
withstand the weight of piping	6								
examine without dismount	6								
utilization as inline pump	6								
reduce recource consumption	9								
small and light	6								
improvement of part yield	6								
less electoricity usage	9				9				
reduction of renewal of lubricating oil	6				6				
less packaging material	6								
easy to install	1								
less noise & vibration	6				1				
less faults	9								
less material for maintenance	6								
spare parts available	6								
less maintenance	6								
easy to service	9								
durable painting	6								
durable	9								
less toxicity	6								
easy to disassemble	6								
	Raw Score	279	225	174	417	15	153	135	153
	Relative Weight	0.042	0.034	0.026	0.063	0.002	0.023	0.021	0.023

Table 2. The table for Phase I of QFDE (cont. from the previous page)

Engineering Metrics (with Units)																			
leak from packing	temperature of liquid	dencity of liquid	viscosity of liquid	vapor pressure	character of liquid	mass	material	volume	physical lifetime	cutting loss	emission of CO_2	emission of VOC	vibration	noise	recycle ratio	disassemble time	landfill volume	toxicity	number of materials
cm^3/h	degree	kg/m^3	$Pa \cdot s$	Pa	pH	kg	—	m^3	h	kg	kg	kg	db	db	%	h	kg	kg PAH-Equiv.	—
1	1	1	1	3	1	1	1			1	9	1	6	6	1				
3	9	9	9	9	3	1	1			1	9	1	6	6	1				
3	3	9	9	3	9	1	1			1	9	1	6	6	1				
1	1	1	1	1	1	1	1			1	9	1	6	6	1				
				9	6				9										
					6								6	6				1	
1											6		6	6					
6											6		6	6					
			6		9														
											1								
					3		9		1									1	
							9		6										
						9		9											
													9	1					
6						9	9			9	6				9	6	9	1	6
						9	9				6					6	1		
							9			9	9	6			9		9		1
											9	6	1	1			6	1	
																6	1	1	
						6	1	9			9				9		9		
						9	6						6	1					
						1	6						9	9					
6							6								0		0		
6							9			6	6	6			9	6	9		1
3							9				1	1			9	6	1		1
							9				6	9			6	6	9	1	1
						6		6			1				6	9	1		1
1							9					9			1		0	6	
1															6	1	9		
							1		6		1	6			9	1	6	9	1
						1	6	6		1	1				6		1		
256	22	28	34	67	82	312	696	198	108	189	579	288	309	256	549	330	501	130	93
0.039	0.003	0.004	0.005	0.010	0.012	0.047	0.106	0.030	0.016	0.029	0.088	0.044	0.047	0.039	0.083	0.050	0.076	0.020	0.014

Table 3. The table for Phase II of QFDE (to be cont. on the next page)

QFD for Environment Phase II			capacity	total head	rotating speed	efficiency	specific speed	suction pressure	discharge pressure	NPSH (Net Positive Suction Head)	leak from packing	temperature of liquid
Part Characteristics (with Units)			0.042	0.034	0.026	0.063	0.002	0.023	0.021	0.023	0.039	0.003
casing	stage	—	1	9		1		3	3			
	suction bore dia	mm	9	3		1	6	9	3	9		
	material	—										3
impeller	impeller diameter	mm	3	9	9	3	9	3	3	3		
	number of blades	—	3	3	3	3	3			3		
	material	—	1	1	6	3		1	1	1		3
shaft	diameter	mm					1				1	
	rigidity	N/m^3	1	1	1							
	material	—	1	1	1							6
bearing	amplitude	1/1000 mm										3
	material	—										6
packing	material	—									9	6
base	installed space	m^2	1									
motor	rotating speed	rpm			9	3	9					
	thermal class	—										
	output power	kW	1	1	1	1						
	phase voltage	V										

3.2.3 Phase II

Table 3 shows the results from Phase II of QFDE. The results indicate that "casing material", "packing material", "impeller material", "impeller diameter", and "casing stage" are the most important PC (part characteristics) in the order. It was found that materials of the parts have relatively great influence on the VOC and VOE. Note that the shaded part is environmental here, too. In addition, PC were specified by parameters of a component in this case, since these can produce more implications than only PC.

3.2.4 Phase III and Phase IV

Based on the results from Phase I and II and technical feasibility, the QFDE team proposed an ecodesign option to reduce the casing weight in Phase III. According to it, Phase IV revealed that the redesign option would contribute

Table 3. The table for Phase II of QFDE (cont. from the previous page)

Engineering Metrics (with Relative Weights)																		Raw Score	Relative Weight
dencity of liquid	viscosity of liquid	vapor pressure	character of liquid	mass	material	volume	physical lifetime	cutting loss	emission of CO_2	emission of VOC	vibration	noise	recycle ratio	disassemble time	landfill volume	toxicity	number of materials		
0.004	0.005	0.010	0.012	0.047	0.106	0.030	0.016	0.029	0.088	0.044	0.047	0.039	0.083	0.050	0.076	0.020	0.014		
				9	3	9		6					3	9	3		3	2.71	0.081
		9				6							1	1	0			1.45	0.044
1	1		9	3	9	6	6	9					9	3	6		6	3.21	0.096
6	3	1	3	3	3	9		9					3	3	6		3	3.06	0.092
1	1	3	3	3	3			9					3	3	3			2.00	0.060
			9	3	9		6	6					9	1	3	3	6	3.15	0.095
				6	6	6		3							1			1.30	0.039
				3	9			3							0		3	1.33	0.040
			9	6	9		6						6	1	3		6	2.43	0.073
	3		3		3		6				9	9						1.25	0.038
			9	1	9		6							1	3	3		1.57	0.047
6	6		9	1	9		9						9	1	6	9	3	3.17	0.095
				6	3	9					1			3	0		1	1.13	0.034
				6					9	9	1	9						2.32	0.070
				6	6									1	0			0.97	0.029
3	3			6					9	9	1	1		1	0			1.80	0.054
									3	3								0.40	0.012

to improve the following customer requirements. Note that the numbers in the parentheses are "improvement rate" calculated in QFDE Phase IV.

- examine without dismount (0.097)
- easy to install (0.079)
- small and light (0.056)
- easy to service (0.039)

The calculation procedures are briefly explained. First, the "improvement rate" for the target EM, "mass", was calculated to be 0.19 through dividing the sum of the scores in the intersections between "casing" and "mass" in table of Phase II ($9+0+3=12$) by the sum of all the scores in the column "mass" ($9+0+3+3+3+3+6+3+6+0+1+1+6+6+6+6+0=62$). Then, for instance, the "improvement rate" for "examine without dismount" was calculated to be 0.097 through dividing the product of 0.19 and the score in the intersections between "mass" and "examine without dismount" in table of Phase I (9) by the sum of all the scores in the line "examine without dismount" ($9+9=18$).

It was found that the weight reduction of the casing contributes to fewer loads of on-site activities by engineers after distribution. This originates from the property that the pump is quite a heavy product. In addition, "make it smaller and lighter" is highly influenced. However "reduce resource consumption" is not so affected, since "reduce resource consumption" is scored to be affected by many other EM than "mass" in QFDE Phase I. This reason is in reality not always theoretically true, however, this is a limitation in QFD and QFDE.

3.3 LCA of the pump

3.3.1 The LCA method

The process of LCA is standardized by ISO [2]. It requires quantitative information on the environment of a product, its manufacturing, and so on. Once a calculation model of LCA is constructed, the influences by a change of a parameter value, for instance, on an impact category, can be calculated, which corresponds to sensitivity analysis.

3.3.2 The scope

The system boundary in this study is set to be from material production, assembly, usage, recycle, to disposal of the pump. Recycled materials are input again into material production stage. The recycle ratio of the steel is assumed to be 100%, while the other metals 50%.

3.3.3 Inventory analysis and impact assessment

The inventory analysis employed Gabi Ver.3 [17]. In the impact assessment part, the classification and characterization were carried out based on the data of Gabi Ver.3, while the weighting adopted the Eco Indicator 95 method [18]. Produced results include that about 98% of the total global warming potential comes from its usage phase.

In order to prepare for the comparison to be presented in Chapter 4, a value in each impact category was allocated to the parts, although this is not normally achieved in LCA. Note that it is impossible to allocate an environmental burden or impact to each part, because it sometimes arises from an activity in which multiple parts participate. For instance, energy usage in the assembly, usage, and disassembly cannot be split to each part's contribution. Thus, environmental burdens in the material production stage including exemption by recycling were allocated to the parts based on the constituent materials. The environmental burdens in the assembly and disassembly were not counted because they were negligibly small. Those in usage stage were not allocated since no appropriate method is found.

As a result, the contribution rate of the casing on energy usage was calculated to be 40% through dividing the amount of energy usage originating from the casing (1,900 MJ) by that from all the parts (4,800 MJ). See the results for all the impact categories in Table 5 in Chapter 4.

3.3.4 Sensitivity analysis

A simple sensitivity analysis was achieved in case that the casing weight was reduced by 10%, so that the effects on the impact categories were investigated. Note that the lightened part corresponds to the target part in Phase III of QFDE described in Section 3.2.4. "Sensitivity" was calculated by the equation below:

$$Sensitivity_n = \frac{I_{n,original} - I_{n,reduced}}{I_{n,original}} \Big/ 0.1$$

where $I_{n,original}$ and $I_{n,reduced}$ mean that a score for impact category n of the original product and of the lightened product, respectively. The bigger sensitivity means that the reduction on the casing weight would affect the greater on the impact category. Table 4 shows a sensitivity for each impact category.

Table 4. The results from sensitivity analysis

Impact categories	Sensitivity
Acidification potential	6.4%
Aquatic ecotoxicity potential	34.7%
Carcinogenic sunstances	17.4%
Eutrification potential	34.5%
Global warming potential	0.2%
Humantoxicity potential	31.1%
Photochemical oxidant potential	3.2%
Terrestric ecotoxicity potential	30.1%
Winter smog	3.6%

The results imply reducing the casing weight would contribute to decrease the aquatic ecotoxicity potential, human toxicity potential, and terrestric ecotoxicity potential. This originates from the LCA model, which describes that the weight of the casing is related only in the material production stage and the toxicity potential in that stage is considerable. Though the casing weight difference can affect environmental burden on life cycle stages other than material production, this kind of effect is not calculated because of its absence in the LCA model.

4 Implications from the case study

4.1 Analysis of QFDE and LCA results

4.1.1 Comparison of the important parts

"Importance" of the parts in the product was investigated in the results of QFDE and LCA.

In QFDE, the results of the "relative weights" of part characteristics in Phase II tell relative importance of the parts. In the columns of "Only VOE" and "VOE and VOC" in Table 5, a relative importance of each part characteristic is shown when only the VOE were considered as customer requirements and when both VOE and VOC were considered, respectively. Scores in the two columns do not vary much, however, the scores in the "Only VOE" column should be compared with LCA results.

The results of allocation of environmental impact to each part in LCA are obtained through the procedures explained in Section 3.3.3. The "mass" column shows the mass ratio for each part. Note that the total is not 100%, because other minor parts exist in the product.

Table 5. The comparison of the results from LCA and QFDE

Item / Part characteristics		Results from LCA						Results from QFDE: Only VOE	Results from QFDE: VOE & VOC
	Mass	Energy usage	Global warming	Acidification	Eutrification	Carcinogenic	Winter smog	Relative importance	Relative importance
casing	40%	40%	39%	34%	37%	36%	36%	22%	23%
impeller	3%	11%	11%	6%	0%	0%	18%	23%	26%
shaft	3%	2%	3%	1%	1%	1%	3%	17%	17%
bearing	6%	6%	6%	4%	5%	5%	6%	7%	7%
packing	0%	0%	0%	0%	0%	0%	0%	11%	7%
base	20%	19%	18%	18%	19%	19%	17%	4%	4%
motor	23%	22%	23%	36%	39%	38%	21%	15%	17%

As shown in Table 5, quite a few parts whose relative importance from QFDE are high but not so from LCA, including impeller, shaft, and packing. The first reason is the difference between how to treat life time in the two methods. In QFDE, those three parts are found to be important since they have strong correlations with the physical lifetime. On the other hand, in LCA, the information on which part affects on the product life time has no impact on the "importance" of the parts. The second reason is the difference between the effects of the mass in the two methods. When the two correlation matrices in QFDE are scored, the functionality rather than the mass of a part is considered. On the other hand, the importance of the parts obtained from the results of LCA here strongly depends on their mass because of the adopted allocation method.

"Base" belongs to the category different from the three parts above. Namely, its relative importance from LCA is high but not so from QFDE. This is because the mass is large but the functionality is not regarded as high.

Engineers should pay much attention to the results from QFDE than from LCA in order to make a focus on product (re)design or technology development, because functionality of a part is not considered in the calculation procedures of LCA. It should be noted that functionality is a core concept to drive engineering activities, not attributes of a product, as is widely recognized in the field of engineering design [16, 19]. This does not necessarily mean that the results from LCA are of no use for such purposes due to the subjectivity in the results of QFDE, which originates from the expert judgement utilized to make matrices in QFDE and the arbitrariness to select VOE. The results of LCA are helpful to know the environmental profile of a product as-is and more objectively than those of QFDE, since the procedures of LCA are more objective and quantitative-data intensive.

4.1.2 Comparison of the effects from the weight reduction

Results in the case of casing weight reduction described in Section 3.2.4 and 3.3.4 are compared.

Toxicity impacts, which the LCA indicated would be decreased as shown in Table 4, did not seem influenced factors in QFDE, since a requirement related to toxicity was not correlated with the product weight in QFDE. This comes from the feature that QFDE includes subjective processes and LCA is objective.

On the other hand, a requirement on easy maintenance activities is presented to be affected by reducing the casing weight from QFDE, which LCA cannot imply.

4.2 Characteristics of QFDE and LCA methods

4.2.1 The roles of the parts

A functionality of a part effective on environmental burdens can be represented in QFDE. For example, the packing of the pump, which plays a large role in spite of its small weight, was found to be important in QFDE of the pump. This can be a good implication to engineers ecodesigning a product.

4.2.2 The factors on the lifetime

The factors on the lifetime of a product can be represented in QFDE. For instance, the packing of the pump, which is critical to the lifetime of a pump, was found to be so in the QFDE case. Due to this characteristic, QFDE can provide designers with information on how to realize long lifetime of a product. On the other hand, LCA cannot give the information on which part's modification would lead to lengthening the lifetime of a product.

4.2.3 The effects of maintenances

Activities such as maintenance can be represented as a process in LCA. However, critical factors in those activities cannot be represented, although inputs and outputs of the activities are described. QFDE can represent that type of information. For instance, in QFDE of the pump, it was shown that reducing the casing weight would make maintenance easier. This is also useful for ecodesign engineers.

4.2.4 Objective and quantitative information

Since QFDE includes subjective scoring in its process, this makes it the more probable to miss some aspects of the reality. This was actually found in the casing weight reduction example as described in Section 4.1.2. On the other hand, LCA is objective, thus LCA in principle can avoid the problem above.

5 Conclusions

As a result from the analyses, results from QFDE were proved to output the importance factors of components considering their functional roles from the environmental viewpoint, which serve as fundament in engineering activities. This type of information supports designers effectively but cannot be obtained only from LCA. On the other hand, the results produced by LCA are indispensable for obtaining quantitative and objective environmental performance of a product. Hence, a combinatorial usage of LCA and QFDE is suggested to be useful to identify significant environmental effects of a product and providing designers with the information about priorities on making their product environmentally conscious. Such a combinatorial usage works also for making a focus in technology development.

Acknowledgement. This research is based on a case study conducted under a committee for standardizing ecodesign methods which the Japan Machinery Federation have run since the fiscal year 2001. The authors would like to thank all the members in the committee. This research was partially supported by a Research Fellowship Programme by Alexander von Humboldt Foundation in Germany.

References

1. ISO (2002) ISO/TR 14062, Environmental management — Integrating environmental aspects into product design and development.
2. ISO (1997) ISO 14040, Environmental management — Life cycle assessment — Principles and framework.
3. Masui, K., Sakao, T., Kobayashi, M. and Inaba, A. (2003) Applying Quality Function Deployment to environmentally conscious design. Intl. J. Quality & Reliability Management, 20 (1) 90–106.
4. Akao, Y. (1990) Quality Function Deployment. Productivity Press, Cambridge, Massachusetts.
5. Ernzer, M., Sakao, T. and Mattheiß, C. (2004) Effectiveness and Efficiency Application to Eco-QFD. Intl. Design Conference, 1521–1526.
6. Sakao, T., Masui, K., Kobayashi, M., Aizawa, S. and Inaba, A. (2001) Quality Function Deployment for Environment: QFDE (2nd Report) — Verifying the Applicability by Two Case Studies —. Intl. Symp. Environmentally Conscious Design and Inverse Manufacturing, 858–863.
7. Fargnoli, M. (2003) The Assessment of the Environmental Sustainability. Intl. Symp. Environmentally Conscious Design and Inverse Manufacturing, IEEE Computer Society, 362–368.
8. IPCC (Intergovernmental Panel on Climate Change) (2001) Climate Change 2001: The Scientific Basis, Cambridge University Press.
9. OECD (2001) OECD Environmental Outlook.
10. DeSimone, L. and Popoff, F. (1997) Eco-efficiency — The Business Link to Sustainable Development —, MIT Press.
11. http://www.meti.go.jp/english/policy/index_environment.html
12. EU (2003) Directive 2002/96/EC of the European Parliament and of the Council of 27 January 2003 on waste electrical and electronic equipment (WEEE).
13. EU (2003) Directive 2002/95/EC of the European Parliament and of the Council on the restriction of the use of certain hazardous substances in electrical and electronic equipment (RoHS).
14. Commission of the European Communities (2003) Proposal for a Directive of the European Parliament and of the Council on establishing a framework for the setting of Eco-design requirements for Energy-Using Products (EuP).
15. Sakao, T., Masui, K., Kobayashi, M. and Inaba, A. (2002) QFDE (Quality Function Deployment for Environment) and LCA: An Effective Combination of Tools for Ecodesign, Intl. Symp. CARE INNOVATION'2002. CD-ROM.
16. Pahl, G. and Beitz, W. (1996) Engineering Design — A Systematic Approach. Springer, London.
17. http://www.ikpgabi.uni-stuttgart.de/english/index_e.html
18. http://www.pre.nl/default.htm
19. Suh, P. (1990) The Principles of Design, Oxford University Press.

Communication Gap Management Towards a Fertile Community

Naohiro Matsumura

Graduate School of Economics, Osaka University
1-7 Machikaneyama, Toyonaka, Osaka, 560-0043, Japan
matumura@econ.osaka-u.ac.jp

Summary. In the paper, we first present an approach to extract social networks from online conversations. Then we propose communication gaps based on structural features of the social networks as an indicator of understanding the state of communication. After we classify 3,000 social networks into three types of communication, i.e., interactive communication, distributed communication, and soapbox communication, we suggest communication gap management to identify the types of communication, the roles of individuals, and important ties, all of which can be a chance towards fertile community.

1 Introduction

With the advent of popular social networking services on the Internet such as Friendster[1] and Orkut[2], social networks are again coming into the limelight with respect to this new communication platform. A social network shows the relationships between individuals in a group or organization where we can observe their social activities. In this paper, we define a "fertile community" as the community where individuals communicate according to their aims, and output innovative outputs. Communication gap management is proposed here to identify the types of communication, the roles of individuals, and important ties, all of which can be used for drawing up a plan for fertile communication.

2 Extracting Social Networks from Online Conversations

In a social network based upon online communication, the distance between individuals does not mean 'geographical distance' because each person lives in a virtual world. Instead, distance can be considered 'psychological distance'

[1] http://www.friendster.com/
[2] http://www.orkut.com/

and this can be measured by the "influence" wielded among the members of the network.

We measure the influnce by using the IDM (Influence Diffusion Model) algorithm in which the influence between a pair of individuals is measured as the sum of propagating terms among them via messages [4]. Here, let a message chain be a series of messages connected by post-reply relationships, and the influence of a message x on a message y (x precedes y) in the same message chain be $i_{x \to y}$. Then, $i_{x \to y}$ is defined as

$$i_{x \to y} = |w_x \cap \cdots \cap w_y|, \tag{1}$$

where w_x and w_y are the set of terms in x and y, respectively, and $|w_x \cap \cdots \cap w_y|$ is the number of terms propagating from x to y via other messages. If x and y are not in the same message chain, we define $i_{x \to y}$ as 0 because the terms in x and y are used in a different context and there is no influence between them.

Based on the influence between messages, we next measure the influence of an individual p on an individual q as the total influence of p's messages on other's messages through q's messages replying to p's messages. Let the set of p's messages be α, the set of q's messages replying to any of α be β, and the message chains starting from a message z be ξ_z. The influence from p onto q, $j_{p \to q}$, is then defined as

$$j_{p \to q} = \sum_{x \in \alpha} \sum_{z \in \beta} \sum_{y \in \xi_z} i_{x \to y}. \tag{2}$$

Here we see the influence of p on q as q's contribution toward the spread of p's messages. The influence of each individual is also measurable using $j_{p \to q}$. Let the influence of p be k_p, and all other individuals be γ. Then, k_p is defined as

$$k_p = \sum_{q \in \gamma} j_{p \to q}. \tag{3}$$

The influence between individuals also shows the distance between them with respect to contextual similarity since the influence indicates the degree of their shared interest represented as terms. The influence and contextual distance between individuals are inversely related; i.e., the greater the influence, the shorter the distance. Here, let us define the length of a link (i.e., distance) as follows.

The distance of a link: The distance from an individual p to an individual q, $d_{p \to q}$, is defined as the value inversely proportionate to the influence from p to q; i.e., $d_{p \to q} = 1/j_{p \to q}$.

The distance is between 0 and 1 when the influence is more than 0. However, the distance cannot be measured by the above definition if the influence is 0. In that case, we define the distance $n - 1$ (n is the number of individuals participating in communication) as the case of the weakest relationships; i.e., the diameter of a social network where all individuals are connected linearly with maximum distance. In this way, an asymmetric social network with distance is extracted from online conversations.

3 Communication Gaps in Social Networks

Based on the forward and backward distances between two individuals in a social network, we can consider structural features between them as "communication gaps" where we can understand the state of their communication. In this paper, we consider the distance of the shortest path as the distance between two individuals since it reflects the real channel of their communication. We can also consider another approach to identifying the state of communication by using only the distances between individuals instead of the differences between distances since the distance itself shows another aspect of communication gaps. The approach is known as "closeness centrality" where individuals nearby are more like to give/receive information more quickly than others [2].

Based on the shortest path and closeness centrality, we propose five indices for measuring the collective communication gap, G_{diff}, G_{max}, G_{min}, C_{diff} and C_{dist}, as follows.

$$G_{diff} = \left(\frac{1}{2}\sum_{p\in\gamma}\sum_{q\in\gamma}|d_{p\to q} - d_{q\to p}|\right)\left(\frac{1}{{}_nC_2(n-1)}\right), \tag{4}$$

$$G_{max} = \left(\frac{1}{2}\sum_{p\in\gamma}\sum_{q\in\gamma}\max(d_{p\to q}, d_{q\to p})\right)\left(\frac{1}{{}_nC_2(n-1)}\right), \tag{5}$$

$$G_{min} = \left(\frac{1}{2}\sum_{p\in\gamma}\sum_{q\in\gamma}\min(d_{p\to q}, d_{q\to p})\right)\left(\frac{1}{{}_nC_2(n-1)}\right), \tag{6}$$

$$C_{diff} = \left(\sum_{p\in\gamma}|c_p^{in} - c_p^{out}|\right)\left(\frac{1}{n(n-1)^2}\right), \tag{7}$$

$$C_{dist} = \left(\sum_{p\in\gamma}c_p^{out}\right)\left(\frac{1}{n(n-1)^2}\right), \tag{8}$$

where the distance of the shortest path from an individual p to another individual q is $d_{p\to q}$ ($d_{p\to q} = 0$ if p is equal to q), the set of all individuals in a social network is γ, $|d_{p\to q} - d_{q\to p}|$ is the absolute value of $(d_{p\to q} - d_{q\to p})$, $\max(d_{p\to q}, d_{q\to p})$ returns the maximum value from $\{d_{p\to q}, d_{q\to p}\}$, $\min(d_{p\to q}, d_{q\to p})$ returns the minimum value from $\{d_{p\to q}, d_{q\to p}\}$, c_p^{in} means the inward closeness centrality that shows the sum of distances of the shortest paths from all other individuals to p, c_p^{out} means the outward closeness centrality that shows the sum of distances of the shortest paths from p to all other individuals, ${}_nC_2(n-1)$ is a theoretical maximum of the first term used for normalizing G_{diff}, G_{max}, and G_{min}, and $n(n-1)^2$ is a theoretical maximum of the first term used for normalizing C_{diff} and C_{dist}.

4 Three Types of Communication

To determine the features of the five indices proposed in Section 3, we analyzed 3,000 social networks. The analysis procedure was as follows.

Step 1. We downloaded 3,000 message boards from 15 categories of Yahoo!Japan Message Boards. To equalize the number of messages for each message board, we selected message boards having more than 300 messages and downloaded the first 300 messages. Then, we removed stop words (words except for noun and verb words) from all the messages to accurately measure content-derived influence. In this way, we prepared 3,000 message boards with each having 300 messages.
Step 2. We extracted a social network from each message board using the approach described in Section 2. To equalize the number of individuals in a social network, we constructed a social network with the 10 most influential individuals identified by Equation (3). We thus obtained 3,000 social networks, each consisting of 10 individuals.
Step 3. We measured G_{diff}, G_{max}, G_{min}, C_{dist}, and C_{diff} for the 3,000 extracted social networks.
Step 4. The average of each index was calculated for each category.

The values of the five indices for the fifteen categories are shown in Table 1. Here, to investigate the relationships between the indices and categories, we applied hierarchical cluster analysis to the data. This analysis merges clusters based on the mean Euclidean distance between the elements of each cluster [1]. A tree-like diagram, called a dendrogram, is then constructed as shown in Figure 1. From this figure, we can find three major clusters, each corresponding to a type of communication. We named the clusters as follows.

Interactive Communication: This cluster includes seven categories ("Arts", "Sciences", "Health & Wellness", "Culture & Community", "Romance & Relationships", "Hobbies & Crafts", and "Family & Home"), the indices of which are considerably smaller than those of other categories. The topics in these categories are common and familiar to many individuals who share these interests. As a consequence, individuals are naturally involved in the communication, and actively exchange their ideas with others.
Distributed Communication: This cluster includes six categories ("Regional", "Entertainment", "Business & Finance", "Schools & Education", "Government & Politics", and "Recreation & Sports"), the indices of which are

Table 1. The average of five indices for 15 categories measured from 3,000 message boards in Yahoo!Japan Message Boards.

Categories	C_{dist}	C_{diff}	G_{diff}	G_{max}	G_{min}
Family & Home	0.032	0.028	0.032	0.048	0.017
Health & Wellness	0.065	0.065	0.073	0.100	0.029
Arts	0.068	0.070	0.081	0.108	0.029
Science	0.072	0.068	0.078	0.110	0.034
Cultures & Community	0.085	0.061	0.071	0.120	0.051
Romance & Relationships	0.081	0.079	0.089	0.125	0.038
Hobbies & Crafts	0.095	0.092	0.106	0.146	0.043
Regional	0.161	0.120	0.142	0.230	0.093
Entertainment	0.151	0.129	0.157	0.228	0.075
Government & Politics	0.217	0.167	0.197	0.313	0.120
Business & Finance	0.241	0.160	0.184	0.331	0.150
Schools & Education	0.253	0.161	0.195	0.349	0.158
Recreation & Sports	0.239	0.208	0.253	0.362	0.116
Computers & Internet	0.447	0.221	0.272	0.579	0.315
Current Events	0.455	0.220	0.271	0.588	0.322

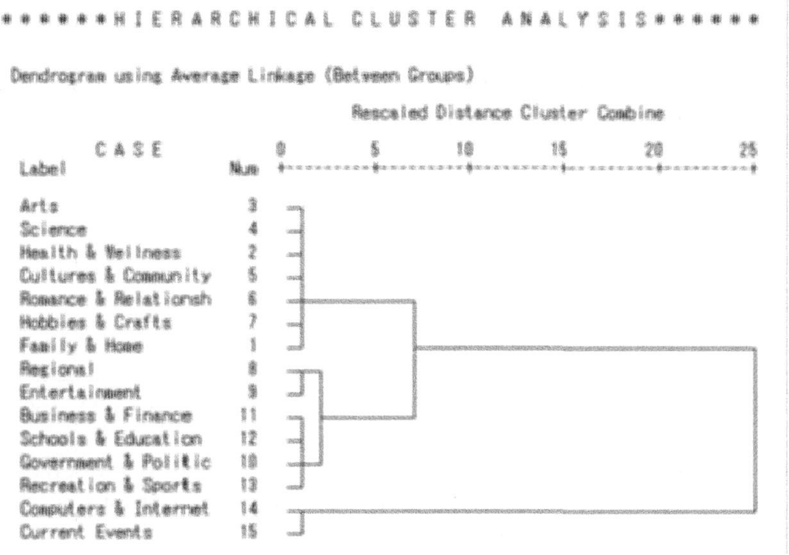

Fig. 1. A dendrogram produced through hierarchical cluster analysis.

generally higher than those of the interactive communication categories. As the topics in these categories are somewhat specific and disputable, experienced or knowledgeable individuals contribute most to the communication.

Soapbox Communication: This cluster includes two categories ("Computers & Internet" and "Current Events"), the indices of which are higher than those of the above two clusters. The topics in these categories are mainly current affairs or topical news, and the communication is one-way from informers to audiences (or lurkers).

From the above results, we can say that there are roughly three types of communication in Yahoo!Japan Message Boards. If these types are common properties in other social networks, it will be possible to identify the state of communication by measuring the five indices.

We expected that the five indices would reveal different aspects of communication, however the Pearson correlation coefficients between them were over 0.9. This meant that these indices were not statistically distinct. In other words, we should be able to identify the types of communication by using only one index instead of all five. In the following, we show some approaches to communication gap management based on G_{max} because G_{max} proved to be the most discriminative index for identifying clusters.

5 Communication Gap Management

Communication gap management is done to identify the types of communication, understand the roles of individuals, and explore remedies for communication gaps in social networks. In this section, we present case studies of communication gap management based on communication types and G_{max} explained before.

5.1 Communication Types

We prepared two types of message log, log 1 and log 2, each of which was extracted from a message board on DISCUS[3]. Log 1 consisted of 72 messages posted by five individuals (one Japanese researcher, one Spanish researcher, and three Japanese businesspersons) who discussed "cell phones and women" in English. Log 2 consisted of 102 messages posted by six individuals (two Japanese researchers, one Spanish researcher, one Japanese businessperson, one Chinese student, and one Taiwanese student) who discussed "purchase of a house" in English. Note that the real names of individuals have been replaced with fictional names to protect their privacy.

Figures 2 and 3 are social networks showing the distance extracted from log 1 and log 2, respectively, using the approach described in Section 2. Once we obtained the social networks, G_{max} was measured as 0.069 from Figure 2 and 0.364 from Figure 3. Comparing G_{max} with the clustering results in Section 4, we can identify the types of communication of log 1 and log 2 as "interactive communication" and "distributed communication", respectively.

[3] http://www-discus.ge.uiuc.edu/

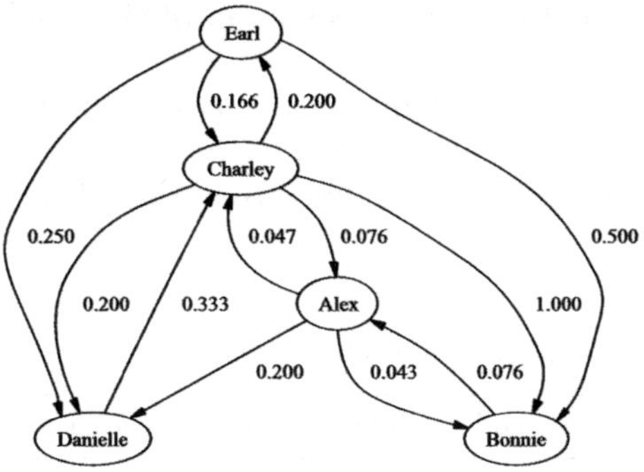

Fig. 2. The social network with distance extracted from log 1. $G_{max} = 0.069$.

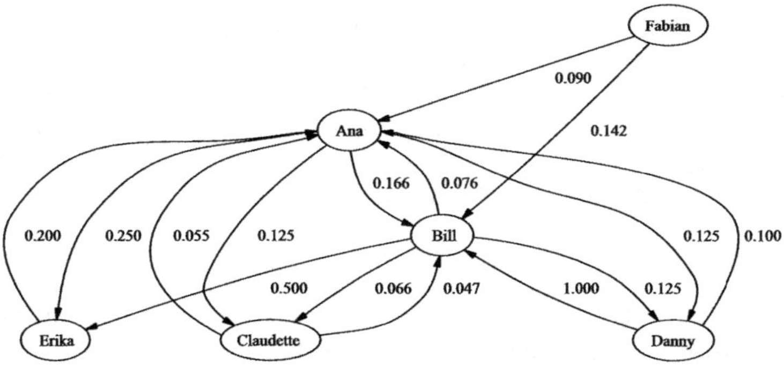

Fig. 3. The social network with distance extracted from log 2. $G_{max} = 0.364$.

The topic in log 1 was familiar to the individuals because they use cell phones everyday. On the other hand, the purchase of a house is a big event in life, and many individuals have no experience of making such a purchase. Therefore, the communication in log 2 was controlled by a few experienced individuals. Thus, the communication types of log 1 and log 2 seem to have properly reflected the types of real communication.

5.2 Roles of Individuals

As shown by the definition of G_{max} in Equation (5), G_{max} is measured by summing each individual's communication gaps with respect to other individuals. That is, G_{max} for each individual is easily measurable by translating the definition of G_{max}. Let the communication gap of an individual p be g_{max}. We can then define g_{max} as

$$g_{max} = \left(\frac{1}{2} \sum_{q \in \gamma} \max(d_{p \to q}, d_{q \to p}) \right) \left(\frac{2}{n(n-1)} \right), \qquad (9)$$

where $n(n-1)/2$ is the theoretical maximum of the first term used for normalizing g_{max}.

As the G_{max} of Figure 3 is high, let us measure each individual's g_{max} to reveal the source of the communication gap. As shown in Table 2, Fabian has the highest g_{max}. From this, we can understand that there are communication gaps around Fabian.

Table 2. g_{max} of each individual in Figure 3.

	Ana	Bill	Claudette	Danny	Erika	Fabian
g_{max}	0.038	0.039	0.038	0.040	0.042	0.167

The roles of individuals are also identified from their relationships with others. If removing a link raises G_{max}, the link is considered to help reduce the communication gap and is therefore important. That is, we can measure the importance of each link by comparing the G_{max} of the original social network with that link to G_{max} of a social network without the link. Let the importance of a link be l_{max}. We then define l_{max} as

$$l_{max} = G'_{max} - G_{max}, \tag{10}$$

where G'_{max} is the G_{max} measured from a social network without the link. From the top three l_{max} values in Figure 3 listed in Table 3, we can see that a link from Erika to Anna is the most important.

Table 3. Three most important links in Figure 3.

Link	l_{max}	G'_{max}	G_{max}
Erika → Ana	0.574	0.938	0.364
Danny → Ana	0.270	0.634	0.364
Ana → Erika	0.226	0.590	0.364

During face-to-face interviews with participants in Figure 3 with showing Figure 3, we found that they considered Fabian creative, but strong-willed to the degree that nobody could counter his ideas. As a result, the communication around him did not go well. They also agreed with the results regarding the important links in Table 3 because these were the links actively used to exchange ideas during the period under study.

5.3 Remedies for Communication Gaps

Once we can obtain information about who are the causes of communication gaps and which links are most important for communication, we can prepare a remedy to improve communication. For example, the following three approaches could be the candidates of the remedies, apart from the feasibility.

- Persuade inactive individuals to contribute to the communication, or persuade strong individuals to listen to others' ideas. While this approach is straightforward, the effectiveness of such persuasion depends on many factors which are beyond the scope of this paper.
- Take inactive or strong individuals away from the communication. This approach is easy and has an immediate effect, but it is not constructive since we would lose the potential contributions of the excluded individuals. For example, Fabian in Figure 3 should not be taken away because he supplied creative ideas although there were large gaps around him. Instead, we should consider how to get him into the communication.
- Add another individual who can communicate with inactive or strong individuals to bridge communication gaps. To find such individuals, the link importance can be used. For example, if we add Mr. X who has a link of distance 1.0 to Anne and from Fabian to the social network in Figure 3, G_{max} would dynamically drop from 0.364 to 0.157.

Needless to say, there approaches above should be planned carefully before putting into practice. Having an inverview with individuals for surveying the effect of a remedy would further enhance the possibility of the success of communicatin gap management. Each remedy mentioned above corresponds to, in other words, a "scenario" for realizing fertile community. As the scenario is analytically drawn up from observed communication, it enables us to realize prompt and smooth revolution of communitation.

6 Related Works

Social networks have been studied for decades. Freeman proposed centrality measures to identify the importance of individuals and ties in a social network [2]. Scott used social networks to identify gaps in information flow within an organization to find ways to get work done more effectively [7]. Krackhardt studied the importance of informal networks in organizations and revealed the effect of these networks on the accomplishment of tasks [3].

Social networks in online communication have been studied as well. Ohsawa et al. classified communities into six types according to the structural features of the word co-occurrence structure of communications [6]. Tyler et al. analyzed e-mail logs within an organization to identify communities of practice — informal collaborative networks — and leaders within the communities [8]. Matsumura et al. revealed the effect of anonymity and ASCII art on communication through message boards [5].

7 Conclusion

Human beings are social creatures, and we could not survive without cooperating with others. It also suggests that understanding how relationships are created and functioned is essential to make our lives happier and richer. The range of our social networks is rapidly expanding than before as the Internet accelerates our communication via E-mail, Chat, Video conference system etc. Reflecting such a situation, managing and utilizing social networks will be an increasingly important part of our lives. We hope this study will contribute to the realization of a better way of life through human relationships over the Internet as well as real world.

References

1. Everitt, B., Landau, S. and Leese, M. (2001) *Cluster Analysis*, Hodder & Stoughton Educational.
2. Freeman, L.C. (1978) Centrality in Social Networks, *Social Networks*, **1**, 215–39.
3. Krackhardt, D. and Hanson, J.R. (1993) Informal Networks: The Company behind the Chart, *Harvard Business Review*, July-August 1993, 104–111.
4. Matsumura, N. (2003) Topic Diffusion in a Community, Yukio Ohsawa and Peter McBurney (Eds.), *Chance Discovery*, Springer Verlag, 84–97.
5. Matsumura, N., Miura, A., Shibanai, Y., Ohsawa, Y. and Nishida, T. (2005) The Dynamism of 2channel, *Journal of AI & Society*, Springer, **19**, 84–92.
6. Ohsawa, Y., Soma, H., Matsuo, Y., Matsumura, N. and Usui, M. (2002) Featuring Web Communities Based on Word Co-occurrence Structure of Communications, Proc. of WWW2002, 736–742.
7. Scott, W.R. (1992) *Organizations: Rational, Natural, and Open Systems*, Prentice-Hall, Inc., Englewood Cliffs, New Jersey.
8. Tyler, J.R., Wilkinson, D.M. and Huberman, B.A. (2003) Email as spectroscopy: Automated Discovery of Community Structure within Organizations. (Preprint http://www.lanl.gov/arXiv:cond-mat/0303264)

Verification of Process Layout CAE System *TPS-LAS* at Toyota

Hirohisa Sakai[1] and Kakuro Amasaka[2]

[1] Toyota Motor Corporation
1, Motomachi, Toyota-shi, Aichi-ken, 471-8573, Japan
E-mail: h_sakai@mail.toyota.co.jp
Tel:+81-(565)71-4803

[2] Aoyama Gakuin University
5-10-1 Fuchinobe, Sagamihara-shi, Kanagawa-ken, 229-8558, Japan
E-mail: kakuro_amasaka@ise.aoyama.ac.jp
Tel:+81-(42)759-6313

Summary. Recent Japanese enterprises have been promoting "global production" in order to realize "uniform quality worldwide and production at optimal locations" in an environment of severe competition. In recent years, digital engineering and computer simulation concepts have expanded to include, not only product quality, but also the quality of business processes and company management, while the scope of business administration activities has been widened. Considering the production environment surrounding the manufacturing enterprise, the authors have established a strategic manufacturing technology for Lean Production called *TPS-LAS* (Toyota Production System–Process Layout Analysis Simulation) by using Process Layout CAE System. At this stage of the investigation, this paper concentrates on verifying the effectiveness of the authors' proposed *TPS-LAS* in Toyota's manufacturing technology.

Key words: global production, lean production, digital engineering, computer simulation.

1 Introduction

Recent Japanese enterprises have been promoting "global production" to realize "uniform quality worldwide and production at optimal locations" in order to survive in an environment of severe competition. In the future, there will be a need to reform production workshops by transforming production processes and facilities, and to improve quality assurance by utilizing digital engineering and computer simulation. By using these key technologies

* Contributed paper: received date: 3-May-05, accepted: 15-Sep-05

to ensure uniform quality levels worldwide, with production taking place at optimum global production locations, the authors have established a strategic manufacturing technology for lean production. This technology, called *TPS-LAS* (Toyota Production System–Process Layout Analysis Simulation), employs a process layout CAE system. The *TPS-LAS* model contains three-core systems: Logistics investigation simulation (*-LIS*), digital factory simulation (*-DFS*) and workability investigation simulation (*-WIS*). The authors have applied the *TPS-LAS* model at Toyota in order to align optimum levels of productivity, workability and cost, as well as quality when commencing global production systems at both domestic and overseas plants. This paper concentrates on verifying the effectiveness of the *TPS-LAS* model at Toyota [1, 2].

2 Proposal of *TPS-LAS*, process layout CAE system

2.1 Necessity of digital engineering and computer simulation in the manufacture of motor vehicles

Digital engineering and computer simulation are generally interpreted as processes that use information technology to handle three-dimensional data. These technologies are used as tools to innovate processes concurrently by allowing for the visualization of latent problems in conventional systems and processes. The early visualization of latent problems allows systems and processes to be simplified in order to achieve improved efficiency [6, 7].

In this paper, the authors attempt to describe how digital engineering and computer simulation are applied at Toyota. Table 1 shows the evolution of Toyota's use of digital engineering.

Table 1. Evolution of digital engineering at Toyota.

	1996∼	1998∼	2000∼
Subject	Vehicle and product investigation	Facility investigation	Overseas support and plant operation
Objective	Realization of efficient product and prototype development (to reduce lead-time)	Realization of efficient planning and construction of facilities (to reduce lead-time)	Realization of efficient overseas planning and construction of facilities (to minimize the need for language)
Results	· Realization of computer-aided prototype-less processes · Problem solution in design stage	· Realization of lean production system for new body line through advanced investigation of virtual facilities and operation	· Realization of lean production system for assembly line through advanced simulation of virtual facilities and operation
Digital engineering	· Integrated CAD (Wire, surface, solid)	· Digital assembly · Facility investigation	· Manufacturing and line simulation

First, in 1996 digital engineering was applied to the investigation of vehicles, products and production facilities, contributing to epoch-making reductions in lead-time, from product development to the production preparation processes. Next, in 1998 these technologies began to be applied in vehicle development. The result was the realization of a computer-supported process that enabled Toyota to detect and solve problems on the drawing board, without the use of prototypes. In production facilities, digital engineering contributed to cost reductions in the manufacture of stamping dies and to the realization of a lean production system for new body welding lines, etc.

Finally, in the period since 2000, Toyota has put digital engineering technologies into even wider practice, combining and integrated them in order to succeed in global production. These tool yield engineering data that can be presented visually in order to realize the efficient planning and construction of overseas facilities with a minimum need for linguistic communication. There is a particular need for visual data that meets the different requirements of different locations all over the world with respect to assembly lines where human labor is required.

The application of digital engineering provides the following benefits:

(1) Easily understandable visual information;
(2) Information sharing in real-time;
(3) Transformation of implicit knowledge into explicit knowledge (via the proper feedback of high-level manufacturing skill to the production design, etc.)

In contrast to digital engineering, computer simulation has not been applied in practice. It has been used, however, to establish uniform manufacturing technologies for global production processes and facilities. Table 2 shows the evolution of computer simulation concerning *kanban* in particular. The table presents two types of models (ordering quantity model and ordering point of time model), and their systems (pull system, push system, SLAMII, CONWIP) [8, 9, 10, 11, 12, 13]. These models have not been applied because actual line processes do not have the ideal availability and reliability. So it is essential that a practical model for global production be created.

Table 2. Evolution of computer simulation concerning *kanban*

Type	System(1)	System(2)
Ordering quality model	· Pull system · Push system	· Pull-push integration system · Push-pull integration system
Ordering point of time model	· SLAM II	· CONWIP (Constant work-in-process)

2.2 Concept of *TPS-LAS* model

The authors propose *TPS-LAS*, a new principle that implements and integrates digital engineering and computer simulation. *TPS-LAS* consists of the three elements (i), (ii) and (iii) shown in Fig. 1. These elements allow for the following fundamental requirements to be addressed: the renewal of production management systems appropriate for digitized production; the establishment of highly reliable production systems; the creation of attractive workshop environments that allow for a greater number of older and female workers; the renovation of the work environment to enhance intelligent productivity; the development of intelligent operators (improved skill level); and the establishment of an intelligent production operating system.

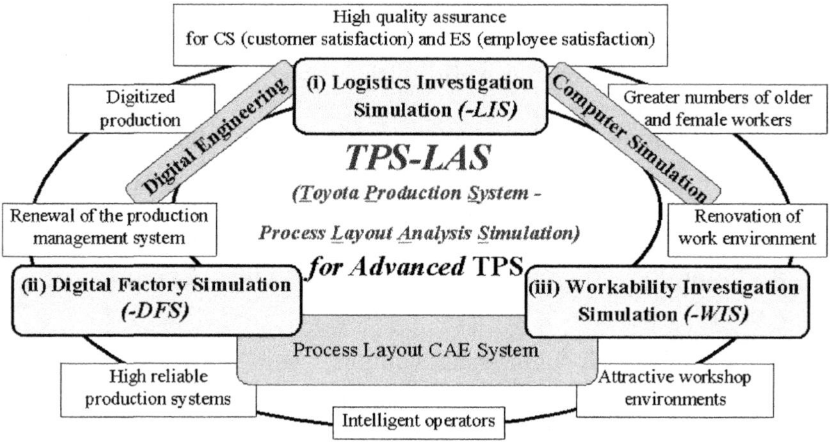

Fig. 1. *TPS-LAS* model using process layout CAE system

The authors have been successful in demonstrating the effectiveness of the three-core systems: (i) logistics investigation simulation [14] (*-LIS*), (ii) digital factory simulation [15, 16] (*-DFS*) and (iii) workability investigation simulation [17] (*-WIS*). The three-core systems are described below.

2.2.1 Logistics investigation simulation (*-LIS*)

Failures such as empty buffer numbers sometimes occur because buffer or transfer numbers between processes in an automobile production line are determined by individual knowledge and the know-how of production engineering personnel. For example, line stoppages resulting from shortages of transfer and career equipment may reduce line productivity. Therefore, an optimum design for buffer numbers is necessary in designing a production system. These buffer numbers are calculated on a theoretical basis. However, because equipment defects can easily cause line stoppages, theoretical calculations have less meaning in actual practice than anticipated.

The authors developed (i) logistics investigation simulation (*-LIS*) to address this problem. Concretely speaking, (*-LIS*) is computer simulation under actual production conditions used to calculate optimum buffer numbers (according to the ratio and interval of equipment defects).

2.2.2 Digital factory simulation (*-DFS*)

When production facilities are used in production processes, the problem of repeat work often arises. For instance, it may not be possible to properly position equipment due to insufficient space between the equipment and the ceiling or because of the interference of piping, etc. In addition to these problems of static interference, problems involving dynamic interference may also occur. For example, when all of the equipment performs the designated cycle for the first time, dynamic interference may occur between different pieces of equipment. In order to minimize repeat work (for example, in the positioning of equipment), it is necessary to confirm not only static interference but also the dynamic interference at an early stage.

Digital factory simulation (ii) (*-DFS*) uses three-dimensional data describing production facilities to simulate dynamic interference. As a system, (*-DFS*) is set up temporarily in the virtual process line by using digital engineering. While the designated operating processes are simulated by computer, technician are able to make interferences and to redesign production processes in advance.

2.2.3 Workability investigation simulation (*-WIS*)

In some manufacturing plants, there are operators who cannot work in the actual production processes due to mismatches between their physical abilities and the demands of the work environment. In order to minimize mismatches between images of the workplace and the actual workplace, workability investigation simulation (iii) (*-WIS*) can be used to present prospective employees with a simulation of an actual work environment (operational procedures and working posture, etc.) and actual workloads.

For example, in the case of removing part of factory products from an operational table, (*-WIS*) can describe the posture of the operator with three-dimensional data, using digital engineering to describe the equipment and operator in a virtual scene. (*-WIS*) can also simultaneously calculate TVAL (Toyota-Verification Assembly Load) to provide a quantitative evaluation of ergonomic workloads [18].

This paper focuses especially on (i), logistics investigation simulation, and outlines the results of applying the process layout CAE system to several applications in order to improve productivity. To be specific, the process layout was designed through a virtual scene on a computer, contributing to epoch-making reductions in lead-time, from product development to production preparation processes. The authors implemented the established *TPS-LAS* model at Toyota and achieved the expected results while also making preparations for implementing the model in Toyota's global production strategy. The process layout CAE system is roughly explained below.

2.3 Process layout CAE system "*TPS-LAS*"

With recent developments in automation and technology, production systems have become extremely large and complicated. This has led to an increase in the number of items that must be addressed when carrying out process planning and line drawing. For example, there have been increases in the number of transfer routes and buffering numbers, as well as in the number of pieces of transfer equipment. These are the parameters that determine line specifications and layout. In designing automobile production lines, buffer numbers are calculated by simulation on a theoretical basis. However, theoretical design is not effective in explaining actual situations such as equipment defects that cause line stoppages. Conventional theoretical methods have a standardized character that is not suitable for actual conditions.

For this reason, the authors developed a process layout CAE system *TPS-LAS*. Production processes based on practical dynamic conditions are confirmed in virtual scenes by using digital engineering and computer simulation. Solutions with high accuracy and the most suitable conditions for the actual circumstances should be selected, no matter how many trial numbers are required to support between theory and practice. Furthermore, there is also the need to plan production processes using concurrent engineering in order to reduce preparation lead-time.

The functions and requirements of the process layout CAE system, particularly those for the assembly line, are shown in Table 3.

Table 3. Functions and requirements for process layout CAE system

Function	Requirement
(1) Quantative analysis of availability/Determination of optimum buffering number	(a) Grasp the real situation of equipment defects and make measurements according to the practical line
(2) Measurement of bottle neck process	(b) Visualize the logistics situation in real time
(3) Implementation of concurrent engineering in process planning	(c) Combine and integrate the conventional CAD/CAM systems

The actual line process, which consists of many robots, a circulatory system of vehicle hangers, and the logistics route, is much too long. The necessary functions in the model are: (1) quantitative analysis of availability based on actual production conditions, and determination of the optimum buffering number between each sub line process: (2) measurement of bottle neck processes, specifically, the surplus or shortage of vehicle hangers at the joint process: and (3) implementation of concurrent engineering in process planning to reduce lead-time, from product development to production preparation processes.

According to these functions, this system is incorporated by (a) using reliability technology to grasp the real situation of equipment defects, depending on the defect model, (b) visualizing the logistics situation in real time and (c) combining conventional CAD/CAM systems and integrating them. The authors, therefore, have perceived the necessity of these functions. They have devised the knowledge and the key to reliability technology [19, 20] concerning (1).

The remainder of this paper focuses particularly on (2) and explains the visualization of the bottleneck process. In order to provide a brief overview of this idea, the authors have described the key element: *a relational model between availability and buffer number.*

First of all, the availability and reliability of an information administration system makes it possible to collect data on the number of equipment defects that occur, and to output the operational availability according to each production process. While the operational availability (or failure ratio, so to speak) is set up randomly, the production number and the buffer number can be calculated in a timely manner. Finally, the authors determined the optimum buffer number based on a calculation of the maximum buffer number.

By sharing the same layout database between the conventional layout CAD system and the new simulation system, process planning and line drawing in concurrent engineering are accomplished with accuracy and efficiency. The parallel processing of each planning and line drawing at the optimal timing improves productivity and allows for efficient planning of production preparation processes.

In addition to (3), the implementation of concurrent engineering in process planning, the system uses IT (information technology) to integrate all functions ((1) through (3)).

3 *TPS-LAS* application

3.1 Results of *TPS-LAS* application at Toyota

An examination of the actual results of *TPS-LAS* applications, proved the efficiency of a main body transfer route at a domestic plant. Fig. 2 shows an example of this route. In order to efficiently design production processes, two steps are required: verification of static interference and verification of dynamic interference.

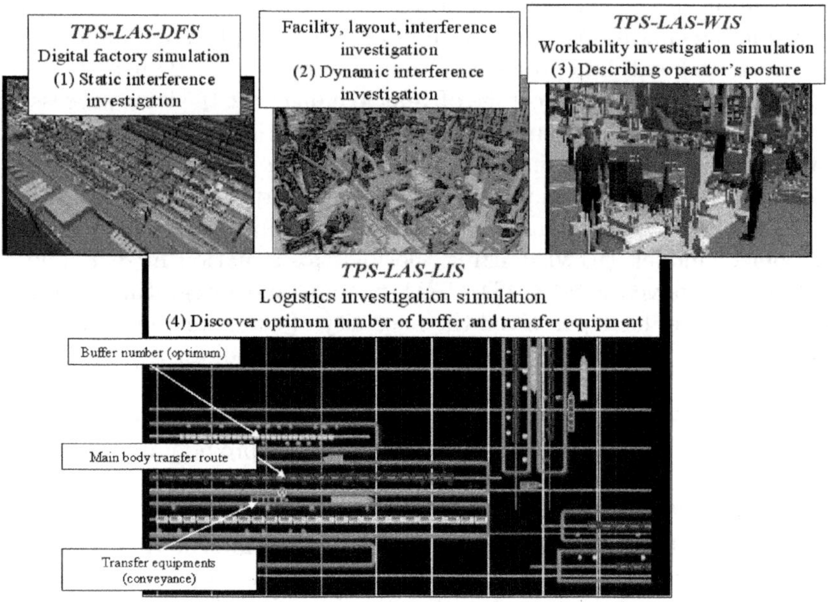

Fig. 2. Result of simulation for the main body transfer route

First, transfer equipment or storage racks are set up temporarily in the virtual process line using three-dimensional data provided by the *TPS-LAS-DFS* digital factory simulation. To begin with product transfer routes and the number of production robots, etc. necessary for the production processes is determined. The production line is then positioned and installed in the plant. Technicians are able to determine the optimum location for the storage racks in each process and to determine the optimum transfer route. At this stage, only static interference is verified. The fundamental requirements for plant construction must be determined at an early stage to prevent critical issues from arising later (such as the need to expand the plant floor during the plant construction stage, etc.) Therefore, in order to allow for the earliest possible examination of these requirements, only static interference is analyzed at this stage, without reference to dynamic information regarding robots and other equipment.

Second, dynamic interference is simulated using, three-dimensional data from all robots and their attachments in the virtual production process. Next, after the dynamic interference in the main structure of the production processes is verified, the dynamic information of each robot is verified by analysis over a long period of time. After that, the dynamic interference of all of the production line equipment is verified by simulating the sequence movement. While the designated operations are simulated, technicians can determine the various interferences between the robots, and between robots and attachments, etc. With this data, the arrangement of robots and attachments can be verified and the production processes can be redesigned at an earlier stage. This allows the position of equipment (for example: the modification of the angle between a robot hand and a manufacturing tool) to be optimized without requiring the repeated physical adjustment of the equipment.

The two-step process of verifying both static and dynamic interference in production engineering design minimizes repeat work throughout the entire production engineering process. In addition, it enables production engineering lead-time to be reduced by narrowing the verification area for dynamic interference.

Third, when extracting part of factory products from the equipment, the products and the operators can be inserted temporarily into the virtual process line using three-dimensional data. The posture of the operator is described by the *TPS-LAS-WIS* workability investigation simulation and verified by TVAL [18] to provide quantitative evaluations of ergonomic workload and workability. When brand-new production systems are installed all over the world, the different laws and regulation governing the work operations and the work environment in each country or each region must be obeyed. For example, there are rules regarding the high and low bending angle of the arms in assembling parts to products, the weight limits for transferring or carrying loads and the forward bending angle. If workloads are higher than the designated level or the operational posture is higher than the designated angle, equipment must be re-design and/or the operator's standing position must be changed. Furthermore, each standardized operation must be proved within the limited production cycle time. Using quantitative analysis, temporary workers can be informed about the actual work environment (operational procedures and working posture, etc.) and workload prior to be hired.

Finally, logistics, the key topic of this section, is investigated by conducting a *TPS-LAS-LIS* logistics investigation simulation. They have looked for the similar production line, applied for the availability and reliability information administration system ARIM-BL [19, 20]. Date on the number of equipment defects that occurred was collected and the failure mode was confirmed. Operational availability was output according to each production process. While the operational availability is set up randomly, the production number and simulated the buffer number will be calculated in a timely way. When the buffer number is empty, indicating a buffer shortages or collisions between pieces of transfer equipment, indicator flash in real time. This enables us to predict the causes of overtime work and declines in assembly line availability. Therefore, by applying this latest method, the authors were able to discover the optimum number of buffers, as well as the optimum number of transfer equipments pieces. This allowed for the development of optimum transfer routes.

Process planning and line drawing in concurrent engineering are accomplished accurately and efficiently by these simulations in advance of planning production preparation processes.

3.2 Effectiveness of TPS-LAS application

The authors' proposed *TPS-LAS* model consists of three-core systems (*TPS-LAS-DFS*, *-WIS*, *-LIS*) and eight fundamental requirements listed on the outside circle. Concretely, the requirements are follows: the renewal of production management systems appropriate for digitized production; the establishment of highly reliable production systems; the creation of attractive workshop environments that allow for a greater number of older and female workers; the renovation of the work environment to enhance intelligent productivity; the development of intelligent operators (improved skill levels); and the establishment of an intelligent production operating system. All of these requirements have been fulfilled. However, among the verified results, the following deserve special mention.

The first core element, *TPS-LAS-DFS*, enabled the authors to accomplish an average eight percent improvement in assembly line availability with digitized and highly reliable production systems. The verification of static and dynamic interference by confirming the positioning of robots, attachments, etc. and the redesign of the production processes at an early stage, reduced repeat work by half.

The second core element, *TPS-LAS-WIS*, enabled the authors to simultaneously verify workloads and workability in advance. This helped to create attractive workshop environments, as well as reform and enhance the work environment. Facilities were re-designed in the following ways: the operator's standing position was modified, and a ten percent improvement in line availability was achieved. Furthermore, voluntary resignations among temporary workers were reduced by using virtual scenes to demonstrate the work environment and workloads prior to hiring.

With respect to overtime work issues, the third core element, *TPS-LAS-LIS*, enabled the causes of overtime work and declines in assembly line availability to be predicted in the process planning and line drawing stage, relating the concurrent engineering as the renewal of production management systems. Therefore, the authors were able to determine the optimum number of buffers and transfer equipment and to develop an optimum transfer route. As a result, the instances of one-hour overtime work was reduced.

TPS-LAS enabled the authors to achieve uniform, high quality assurance at both domestic and overseas plants. It also enabled a reduction of lead-times and the support required for overseas labor forces in global production processes throughout the world. The results of this study have been deployed in Toyota's global production strategy, with the effectiveness of *TPS-LAS* verified by the excellent reputation of recent Toyota vehicles for thier reliability and common workability in Europe and the United States [21].

4 Conclusion

In an environment of worldwide quality competition, establishing a new management technology that contributes to business processes is an urgent goal being sought in the global manufacturing industry. The conventional theoretical simulation model did not apply to the practical design support of production engineering. For example, because actual operation rate do not reach theoretical targets, there are instances in which additional equipments must be added, requiring repeat work.

In this report, the authors have perceived the need for simulation that matches theoretical and actual conditions. For instance, this can be achieved by collecting defect information from the actual equipment directly or by monitoring the actual condition of each piece of equipment with respect to the above issue. In the course of its implementation, the authors also created the process layout CAE system, *TPS-LAS*, as a strategic production quality management model. The system consists of (i) logistics investigation simulation (*-LIS*); (ii) digital factory simulation (*-DFS*); and (iii) workability investigation simulation (*-WIS*). The authors have verified that *TPS-LAS* will contribute to the renovation of management technologies when made organic.

Acknowledgement. The authors would like to thank Mr. Kazutsugu Suita for his help. Mr. Suita is engaged in the vehicle planning & production engineering division at Toyota. This study was partially supported by the Toyota production engineering team.

References

1. Amasaka, K. (2002) New JIT, a new management technology principle at Toyota, *International Journal of Production Economics*, 80, 135–144.
2. Amasaka, K. (2003) Development of Science TQM, a new principle of quality management: Effectiveness of strategic stratified task team at Toyota, *International Journal of Production Research*, 42 (17), 3691–3706.
3. Amasaka, K. and Sakai, H. (2004) TPS-QAS, new production quality management model: Key to New JIT: Toyota's global production strategy, *International Journal of Manufacturing Technology and Management.* (decided to be published, 2006)
4. Sakai, H. and Amasaka, K. (2005) Development of robot control method for curved seal extrusion for high productivity by advanced Toyota production system, *International Journal of Computer Integrated Manufacturing.* (decided to be published, 2006)
5. Amasaka, K. (2005) Applying New JIT — Toyota's global production strategy: Epoch-making innovation in the work environment, *Robotics and Computer-Integrated Manufacturing.* (decided to be published, 2005)
6. Peter, J.S., and David, K.W. (2003) Data visualization in manufacturing decision making, *Journal of Advanced Manufacturing Systems*, 2 (2), 163–185.

7. Christoph, R., Holger, J., Markus, M., and Sabine, B. (2003) Intelligent manufacturing systems project IRMA: Linking virtual reality to simulation, *Journal of Advanced Manufacturing Systems*, 2 (1), 105–110.
8. Kimura, O. and Terada, H. (1981) Design and analysis of pull system, a method of multi-stage production control, *International Journal of Production Research*, 19 (3), 241–253.
9. Takahashi, K., Hiraki, S. and Soshiroda, M. (1994) Pull-push integration in production ordering systems, *International Journal of Production Economics*, 33 (1), 155–161.
10. Hiraki, S., Ishii, K., Takahashi, K. and Muramatsu, R. (1992) Designing a pull-type parts procurement system for international co-operative knockdown production systems, *International Journal of Production Research*, 30 (2), 337–351.
11. Spearman, M.L. (1992) Customer service in pull production systems, *Operation Research*, 40 (2), 948–958.
12. Spearman, M.L., Woodruff, D.L. and Hopp, W.J. (1990) CONWIP: A pull alternative of kanban, *International Journal of Production Research*, 28 (5), 879–894.
13. Tayur, S.R. (1993) Structural properties and heuristics for kanban-controlled serial lines, *Management Science*, 39 (11), 1347–1368.
14. Leo, J.V., Amos, H.C. and Jan, O. (2004) Simulation-based decision support for manufacturing system life cycle management, *Journal of Advanced Manufacturing Systems*, 3 (2), 115–128.
15. Byoungkyu, C., Bumchul, P., and Ho, Y.R. (2004) Virtual factory simulator framework for line prototyping, *Journal of Advanced Manufacturing Systems*, 3 (1), 5–20.
16. Kesavadas, T. and Ernzer, M. (2003) Design of an interactive virtual factory using cell formation methodologies, *Journal of Advanced Manufacturing Systems*, 2 (2), 229–246.
17. Steffen, S., Gunter, S. and Siegmar, H. (2003) Distributed manufacturing simulation as an enabling technology for the digital factory, *Journal of Advanced Manufacturing Systems*, 2 (1), 111–126.
18. Toyota Motor Corporation and Toyota Motor Kyushu Corp. (1994) Development of a new automobile assembly line, *Business Report with Ohkouchi Prize*, 1993 (40th), 377–381. (*in Japanese*)
19. Amasaka, K. and Sakai, H. (1996) Improving the reliability of body assembly line equipment, *International Journal of Reliability, Quality and Safety Engineering*, 3 (1), 11–24.
20. Amasaka, K. and Sakai, H. (1998) Availability and reliability information administration system ARIM-BL by methodology in inline-online SQC, *International Journal of Reliability, Quality and Safety Engineering*, 5 (1), 55–63.
21. JD Power and Associates at, http://www.jdpower.com/ (2004)

Part II

$(DE)^2$: Design of Experiments in Digital Engineering

A Grammar of Design of Experiments in Computer Simulation

Shu Yamada

Graduate School of Business Science
University of Tsukuba, Otsuka 3-29-1, Tokyo 112-0012, Japan
shu@mbaib.gsbs.tsukuba.ac.jp

Summary. Computer simulation plays an important role in modern technology development. In particular, it is used for exploring new methodology, examining developed technologies and so forth. Design of experiments can benefit from computer simulation because it is a series of techniques for performing experiments efficiently and the above-mentioned exploration and examination requires experimental results of simulation. This paper discusses the application of design of experiments to computer simulation as a grammar of design of experiments in computer simulation. Specifically, the grammar described in this paper includes techniques for validating a tentative model of response and factors, screening many factors to identify active factors and approximating the response function by active factors. Traditional approaches to design of experiments are partially useful for these areas. However, some of them cannot be solved by the traditional approaches because of the complexity resulting from computer simulation. Therefore, this paper also adopts a new approach to solve these problems.

Key words: approximation of response function, composite design, fractional factorial design, screening factors, supersaturated design, uniform design

1 Introduction

Computer simulation plays an important role in modern technology development and it is used for validating developing technology, optimizing technology to adapt it for use in the real world and so forth. Another reason for the popularity of computer simulation is the advantages in terms of speed and cost. Therefore, computer simulation is an alternative tool that is increasingly being used in the place of physical experiments. One example is the use an automobile crash simulation instead of carrying out physical experiments. Since physical experiments take time and are expensive to perform, using computer simulation is preferable. Many companies consequently mare promoting the use of computer simulation company-wide because of the advantages.

Physical experiments have been used to develop new technology for a long time. Once a hypothesis regarding a phenomenon has been formed, it is confirmed by physical experiments. Design of experiments is a methodology for conducting experiments effectively; it has been beneficial tool for developing new technology. It originated from the works by Fisher, R.A. at the beginning of 20th century (Fisher (1966)).

Therefore, there is a need to apply design of experiments to computer simulation experiments in addition to physical experiments because of the recent popularity of computer simulations.

This paper discusses an application of design of experiments in computer simulation as a grammar for technology development. Specifically, the grammar described in this paper includes techniques for validating a tentative model involving response and factors, by screening the many factors and approximating the response function using the active factors.

2 Application of design of experiments to computer simulations

2.1 A general approach to using computer simulation for technology development

When computer simulation is applied to technology development in a specific field, the following can be regarded as a general approach.

1. Interpretation of requirements
2. Development of a computer simulation model
3. Application of the developed computer simulation model

At the first step, the requirements for technology development are translated into a technical problem. In other words, the inputs and outputs of the simulation are specified. The second step involves the development of a computer model simulating the real world. The third step involves technology development based on the developed computer simulation model.

2.2 Interpretation of requirements

At the first step, interpretation of requirements, necessary for the development of the technology being considered and other background issues need to be discussed. In other words, the input and output will be specified at this step. After interpreting the requirements, a tentative simulation model can be obtained by applying specialist knowledge of the field.

2.3 Developing a computer simulation model

In order to develop a simulation model, specialist knowledge of the field needs to be applied. Design of experiments provide little assist in this task. Once a tentative simulation model has been established, it is important to verify and validate the model. Verification and validation is sometimes abbreviated as "V & V" in many fields, including computer engineering and systems management. Verification involves comparing the output of the simulation program to the fundamental model of the simulation. It is similar to comparing a computer program to the specifications of the program. At the "Verification" stage, design of experiments may not help so much because the verification activities are mainly based on the model and simulation engineering viewpoints.

However, design of experiments plays an important role in the validation activities that compare the results of the computer simulation with reality. In order to validate a computer simulation model, the simplest approach is to compare the results obtained by the computer simulation with the results obtained from physical experiments. For example, the simplest way to validate a given computer simulation model on the stress of a system is to measure the stress of the system by doing physical experiments and then comparing the results obtained with the computer simulation results. Therefore, design of experiments are helpful in validating a computer simulation model. In practice, there are a plenty of limitations on experiments. Thus, the important issues at this step are taking into consideration the practical limitations in the design of experiments.

2.4 Application of a developed computer simulation model

Once a computer model that is sufficiently accurate has been developed, computer simulation experiments are performed in order to develop new technology. One example is using a computer simulation model to design the fuselage of an airplane. Fuselage design is explored by taking many criteria into consideration, such as air flow, fuselage balance, temperature of the fuselage, and so forth.

The following are two typical situations with which design of experiments helps in the application of a developed computer simulation model. Design of experiments can help with the application of a developed computer simulation by screening many factors and also by approximating the response functions.

A developed computer simulation usually involves a lot of factors that may be impossible to consider at one time. It is easy to imagine that designing a fuselage of an airplane would involve more than 100 factors. While they should all be considered simultaneously, it may not be possible because of the large computation load required. Therefore, it is necessary to screen many candidate factors to find the active factors that affect the response. Using statistical terminology, the task of selecting some active factors out of many candidate factors is called a "screening problem." Design of experiments assists by identifying the active factors efficiently. This type of approximation is not a new requirement in industry. For example, Taguchi (1976) shows an example to obtain factor effects in an integrated circuit design. In this example, the response values are numerically calculated by a model in the field of electronics.

Once some active factors have been identified, the next problem is to approximate the response using the identified active factors in order to make the optimization of the response easier. At this stage, the response function of the active factors is approximated by a simple function in order to find the optimum condition. For example, the stress of a cantilever is simulated by a simple model by using the finite element method. Let us suppose that some active factors have been identified in order to optimize the shape of the cantilever in terms of stress. Based on the identified factors, the stress of the cantilever may be approximated by a simple function, such as a polynomial function of the factors. Design of experiments assists by finding a good approximation and optimizing the response efficiently.

2.5 Grammar of design of experiments in computer simulation

Let us consider the above approach in terms of design of experiments. The above discussion can be summarized as follows:

1. Interpretation of requirements
   Design of experiments' techniques contribute little to this stage.
2. Developing a computer simulation model
   Design of experiments' techniques are useful for validating a developed model by comparing with reality.
3. Applying the developed computer simulation model
   Design of experiments' techniques contribute to screening many factors and finding a useful approximation for the response.

An outline of the above approach is summarized in Figure 1 as a grammar of design of experiments in computer simulation.

In terms of design of experiments, a systematic approach for validating a developed model with reality has not been established yet because of the difficulty of formulating the problem. Let us consider an example that it is necessary to validate a developed computer simulation model with reality by using a few physical experiments as representatives of reality. In such a case, a few experimental points are determined based on technical knowledge of the field. After determining a few experimental points, the physical experimental results and the computer simulation results are compared in order to validate the simulation model. Since physical experimental results contain repeated errors, statistical tests may be helpful when comparing with the computer simulation results. The above example show the difficulty using design of experiments for validating a developed computer simulation model.

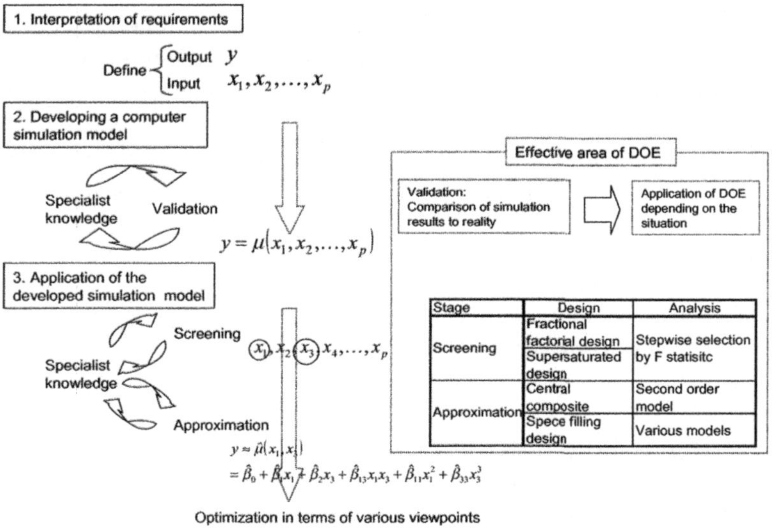

Fig. 1. Outline of a grammar of design of experiments in computer simulation

Other examples that show the difficulty using design of experiments for validating a large-scale computer simulation, include designing an airplane, examining the stress of a building, etc. Since it will be impossible to conduct physical experiments on such a large scale, validation of the computer simulation model has to be performed on a subsystem level. Specifically, the large system has to be decomposed into several subsystems that can be used to validate the subsystem with reality. Although it may be possible to validate a subsystem of the computer simulation model with reality, it will be impossible to validate the computer simulation model as a whole.

While the above examples entail a degree of difficulty in applying design of experiments to validate a developed computer simulation model to reality, there is a need to establish a systematic procedure. In other words, techniques of design of experiments and statistical treatments for validation should be developed in order to accelerate the development of new technology.

On the other hand, design of experiments contributes greatly to both screening of many factors and approximation of responses. Specifically, fractional factorial design and F statistics are helpful for identifying active factors because fractional factorial design can screen more factors than regular factorial design. For approximating the response function, central-composite design and polynomial-function fitting are widely used. A first or second-order model is widely applied for the approximation. Some Computer-Aided Engineering (CAE) software packages have implemented a function that combines composite design and polynomial function fitting by least squares.

While these methods are useful, recent advances in computer simulation require working with more complicated cases for screening and approximation.

For example, Lin (1995) gives an example of computer simulation experiments with more than 100 factors, where it was impossible to apply fractional factorial design because of the limitation on the number of experimental runs that could be performed.

Another example is the application of a high order of polynomial function of factors, which included the 8th power of factors, as described by Fang, Lin, Winker and Zhang (2000). The traditional approach consisting of composite design and fitting using a high order of polynomial function is impossible to fit such a complex function. In the following section, we discuss some advanced approaches for screening and fitting the response that go beyond fractional factorial design and composite design based on a second-order model.

3 Screening many factors

3.1 Problem formulation

Let y be a response variable whose values are calculated by the computer simulation model, and $(x_1, x_2, \ldots, x_p)$ the factors of the response. When the values of the factors $(x_1, x_2, \ldots, x_p)$ are specified, the computer simulation calculates the level of y using the model. Let $\mu(x_1, x_2, \ldots, x_p)$ denote a computer simulation model describing the relationship between the response and its factors. The model $\mu(x_1, x_2, \ldots, x_p)$ is usually a complicated function that can not be described in a simple functional form. For example, the finite element method can be regarded as an example of $\mu(x_1, x_2, \ldots, x_p)$, where the value of $y = \mu(x_1, x_2, \ldots, x_p)$ is calculated when the level of $x_1, x_2, \ldots, x_p$ is specified.

When the number of factors is small, such as 5 or less, it is possible to use $\widehat{\mu}(x_1, x_2, \ldots, x_p)$ as an approximation of the true response function: $\mu(x_1, x_2, \ldots, x_p)$. On the other hand, it is necessary to screen all the candidate factors to find active factors that may affect the response. Once some active factors have been identified, it is necessary to find a good approximating function in order to optimize the response. More details on optimization based on the approximate function are discussed in the following section with an example.

The traditional approach to the screening problem is the application of fractional factorial design and F statistics. For example, fractional factorial design with $16 = 2^{15-11}$ runs can include up to a maximum of 15 factors. The response values are calculated at all experimental settings as determined by the fractional factorial design. After calculating the response values, the F statistic indicates the importance of the factor to the response. A limitation of fractional factorial design is the number of factors that can be included in simulation experiments, *i.e.* the maximum number of factors is $n - 1$ for fractional factorial design, where n is the number of experimental runs. In many practical cases that use computer simulation, this limitation on the number of factors poses a big barrier for screening, because some computer simulations require a long time to perform a calculation.

3.2 An example of screening based on three-level supersaturated design

Supersaturated design

This section describes, using an illustrative example, an approach that involves using many factors in an application of supersaturated design. In this example, supersaturated design replaces the role played by fractional factorial design. Supersaturated design is a type of fractional factorial design that can include more factors than fractional factorial design. It was originally developed by Satterthweitt (1959) as a random balance design and was formulated by Booth and Cox (1962) in a systematic manner. Three decades after its initial development, a constructing method proposed by Lin (1993) triggered recent research into supersaturated designs.

The studies on supersaturated design include construction of designs, mathematical background, application of data analysis, and so forth. Recently, multi-level supersaturated designs have been proposed in many papers, including Yamada and Lin (1999), and Fang, Lin and Ma (2000). Computer simulation experiments sometimes involve many factors; however fractional factorial designs are not able to include many factors due to the limitation on the number of runs. Supersaturated design is a solution to this problem regarding the limitation on the number of runs. For two-level supersaturated design, factors are considered as being active factors on the basis of F statistics, as shown in the example in Lin (1993). Furthermore, performance evaluation of stepwise selection based on F statistics is discussed in Westfall, Young and Lin (1998) and Yamada (2004a) in terms of types I and II errors, respectively.

Two-level supersaturated design has the disadvantage that it is not able to detect higher-order effects. Multi-level supersaturated design was developed for detecting interactions and higher-order effects, making it is useful for computer simulation experiments because there are many factors whose second-order effects are suspected as being active. Tables 1 and 2 give examples of two-level and three-level supersaturated design, respectively.

Outline of the case

An illustrative example of a cantilever-simulation study is used to demonstrate how supersaturated design applies and collected data are analyzed. An outline of the cantilever is shown in Figure 2.

While a physical mechanism has already been obtained, this cantilever example is introduced because its outline can be easily interpreted. The maximum stress of the cantilever is considered as a response variable, and it is calculated using finite element software.

Table 1. An example of two-level supersaturated design by (Lin (1993))

No.	[1]	[2]	[3]	[4]	[5]	[6]	[7]	[8]	[9]	[10]	[11]	[12]	[13]	[14]	[15]	[16]	[17]	[18]	[19]	[20]	[21]
1	1	1	1	1	2	1	2	1	1	2	2	1	1	2	1	2	1	2	2	2	2
2	1	1	1	2	1	2	1	1	2	2	1	1	2	2	2	1	2	2	2	2	1
3	1	1	2	1	2	1	1	2	2	1	1	2	2	1	1	2	2	2	2	1	1
4	1	2	1	2	1	1	2	2	1	1	2	2	1	2	2	2	2	2	1	1	1
5	2	1	2	1	1	2	2	1	1	2	2	1	2	1	2	2	2	1	1	1	1
6	2	1	1	2	2	1	1	2	2	1	2	1	2	2	2	1	1	1	1	1	2
7	1	2	2	1	1	2	2	1	2	1	2	2	2	2	1	1	1	1	2	1	2
8	2	2	1	1	2	2	1	2	1	2	2	2	2	1	1	1	1	2	1	2	1
9	1	2	2	1	2	1	2	2	2	2	1	1	1	1	2	1	2	1	1	2	2
10	2	2	1	2	1	2	2	2	2	1	1	1	1	1	1	2	1	1	2	2	1
11	2	1	2	2	2	2	1	1	1	1	1	2	1	2	1	2	2	1	1	2	2
12	2	2	2	2	1	1	1	1	1	2	1	2	1	1	2	1	1	2	2	1	2

Table 2. Example of three-level supersaturated design (Yamada, *et al.* (1999))

No.	[1]	[2]	[3]	[4]	[5]	[6]	[7]	[8]	[9]	[0]	[1]	[2]	[3]	[4]	[5]	[6]	[7]	[8]	[9]	[0]	[1]	[2]	[3]	[4]	[5]	[6]	[7]	[8]	[9]	[0]	[1]
1	1	1	1	1	1	1	1	1	1	1	1	1	1	1	1	1	1	1	1	1	1	1	1	1	1	1	1	1	1	1	1
2	1	1	2	2	2	3	1	3	2	3	3	2	1	2	3	1	2	3	1	3	2	1	2	2	2	3	1	3	2	3	3
3	1	1	3	3	3	3	3	2	1	2	3	1	2	3	1	3	2	3	3	2	1	1	3	3	3	3	3	2	1	2	3
4	1	2	1	2	3	3	2	1	3	3	1	3	2	3	2	1	3	1	3	3	3	2	1	2	3	3	2	1	3	3	1
5	1	2	2	3	1	2	3	1	2	3	2	1	3	3	3	2	1	3	2	1	3	2	2	3	1	2	3	1	2	3	2
6	1	2	3	1	2	2	2	3	1	2	2	3	1	1	3	3	3	2	3	1	2	2	3	1	2	2	2	3	1	2	2
7	1	3	1	3	2	2	1	2	3	1	3	3	3	2	2	3	1	2	2	3	1	3	1	3	2	2	1	2	3	1	3
8	1	3	2	1	3	1	3	3	3	2	1	2	3	2	1	2	3	2	1	2	3	3	2	1	3	1	3	3	3	2	1
9	1	3	3	2	1	1	2	2	2	1	2	2	2	1	2	2	2	1	2	2	2	3	3	2	1	1	2	2	2	1	2
10	2	1	1	1	1	1	1	1	1	1	1	1	1	1	1	1	1	1	1	1	1	2	2	2	2	2	2	2	2	2	2
11	2	1	2	2	2	3	1	3	2	3	3	2	1	2	3	1	2	3	1	3	2	2	3	3	3	1	2	1	3	1	1
12	2	1	3	3	3	3	3	2	1	2	3	1	2	3	1	3	2	3	3	2	1	2	1	1	1	1	1	3	2	3	1
13	2	2	1	2	3	3	2	1	3	3	1	3	2	3	2	1	3	1	3	3	3	3	2	3	1	1	3	2	1	1	2
14	2	2	2	3	1	2	3	1	2	3	2	1	3	3	3	2	1	3	2	1	3	3	3	1	2	3	1	2	3	1	3
15	2	2	3	1	2	2	2	3	1	2	2	3	1	1	3	3	3	2	3	1	2	3	1	2	3	3	3	1	2	3	3
16	2	3	1	3	2	2	1	2	3	1	3	3	3	2	2	3	1	2	2	3	1	1	2	1	3	3	2	3	1	2	1
17	2	3	2	1	3	1	3	3	3	2	1	2	3	2	1	2	3	2	1	2	3	1	3	2	1	2	1	1	1	3	2
18	2	3	3	2	1	1	2	2	2	1	2	2	2	1	2	2	2	1	2	2	2	1	1	3	2	2	3	3	3	2	3
19	3	1	1	1	1	1	1	1	1	1	1	1	1	1	1	1	1	1	1	1	1	3	3	3	3	3	3	3	3	3	3
20	3	1	2	2	2	3	1	3	2	3	3	2	1	2	3	1	2	3	1	3	2	3	1	1	1	2	3	2	1	2	2
21	3	1	3	3	3	3	3	2	1	2	3	1	2	3	1	3	2	3	3	2	1	3	2	2	2	2	2	1	3	1	2
22	3	2	1	2	3	3	2	1	3	3	1	3	2	3	2	1	3	1	3	3	3	1	3	1	2	2	1	3	2	2	3
23	3	2	2	3	1	2	3	1	2	3	2	1	3	3	3	2	1	3	2	1	3	1	1	2	3	1	2	3	1	2	1
24	3	2	3	1	2	2	2	3	1	2	2	3	1	1	3	3	3	2	3	1	2	1	2	3	1	1	1	2	3	1	1
25	3	3	1	3	2	2	1	2	3	1	3	3	3	2	2	3	1	2	2	3	1	2	3	2	1	1	3	1	2	3	2
26	3	3	2	1	3	1	3	3	3	2	1	2	3	2	1	2	3	2	1	2	3	2	1	3	2	3	2	2	2	1	3
27	3	3	3	2	1	1	2	2	2	1	2	2	2	1	2	2	2	1	2	2	2	2	2	1	3	3	1	1	1	3	1

The factors included in the computer simulation are the widths of the cantilever at different points. The width closest to the fixed wall is denoted by x_1, the width of the next section is denoted by x_2, and so forth. The width of the locating edge of the cantilever is denoted by x_{15}, *i.e.* there are 15 factors in total which represent the width at 15 points of the cantilever. The outline is shown in Figure 2.

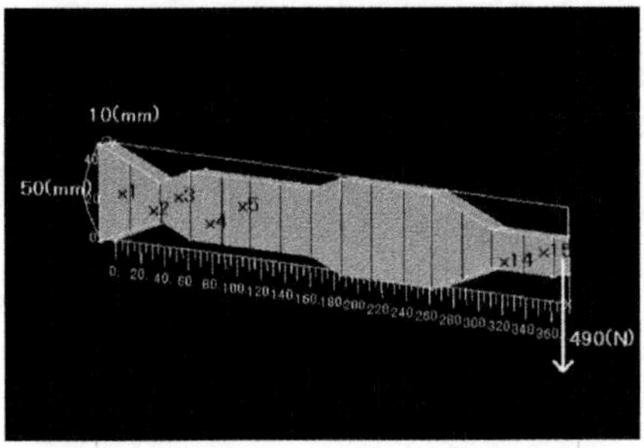

Fig. 2. Outline of cantilever example

Application of supersaturated design

Since it requires a long time to perform the calculation using finite element software, it is necessary to reduce the number of simulation experiments. Furthermore, since the relationship between the response and factors is complicated, three-level supersaturated design is used in order to not only detect linear effects, but also to detect interaction and quadratic effects of the factors. Specifically, $n = 9$-run supersaturated design is applied. An outline of the design is shown in Table 3. We investigate the active factors based on the data shown in this table.

Table 3. Applied three-level supersaturated design and experimental result

No.	x_1	x_2	x_3	x_4	x_5	x_6	x_7	x_8	x_9	x_{10}	x_{11}	x_{12}	x_{13}	x_{14}	x_{15}	y
1	40	40	40	40	40	40	40	40	40	40	40	40	40	40	40	5.76
2	45	40	50	50	45	40	50	50	40	45	50	50	45	40	45	5.22
3	50	45	40	50	40	50	50	45	40	45	50	45	50	50	40	4.82
4	50	45	45	40	50	40	50	50	45	50	40	40	50	45	50	4.42
5	40	50	50	45	40	45	45	40	45	40	50	40	50	50	45	5.37
6	45	40	50	45	50	50	45	40	45	50	45	50	40	45	40	5.22
7	45	50	45	50	50	50	45	50	50	40	45	45	45	40	50	4.49
8	50	50	40	40	45	45	40	45	50	50	45	50	45	50	50	4.93
9	40	45	45	45	45	45	40	45	50	45	40	45	40	45	45	5.48

Screening active factor effects

The number of factors is greater than the number of experimental runs. This fact implies that it is impossible to estimate all the factor effects at a single time. Furthermore, some interaction and quadratic effects may be active. Therefore, stepwise-forward selection based on F statistics is applied to 15 linear effects, $\binom{15}{2} = 105$ interaction effects and 15 quadratic effects. Since it is impossible to estimate all effects simultaneously, stepwise selection is applied to find active factors as well as the data analysis Lin (1993).

For this screening problem, we apply the following principles that are widely recognized for screening based on the F statistic, where the principles are obtained empirically, not mathematically.

Effect sparsity The number of active factors is small, usually less than five.

Effect heredity When the interaction effect of two factors is considered, both of the two linear effects of the two factors should be included in the model.

Order principle Lower order effects are more important than higher order effects.

More details are discussed in many textbooks such as Wu and Hamada (2000).

Based on the above principles, we screen the factor effects by the following procedure.

1. An initial candidate set of effects is constructed using all the linear effects, for example, $\{x_1, x_2, \ldots, x_{15}\}$
2. The effect that has the largest F statistic is treated as an active factor. The quadratic effect of the selected factor is added to the candidate set of factor effects. For example, quadratic effect, $\{x_1^2\}$, is added to the candidate set of the factor effects after selecting x_1. Furthermore, a two-factor interaction effect is added to the candidate set after selecting the two linear effects of x_1 and x_2. For example, the interaction effect of x_1 and x_2 is added to the candidate set after selecting the linear effects of x_1 and x_2
3. The above procedure is repeated until q factors have been selected.

The above procedure ensures that the selected factor effects satisfy the three principles mentioned above. The selected factor effects are x_1, x_1^2, x_7, x_6, x_{15} and x_{15}^2. The three principles are satisfied by these factors. From a mechanical engineering perspective, it is widely recognized that the factors close to the fixed side are active. This confirms the results of screening for the active factors.

3.3 Statistical inference by computer simulation

One of the crucial differences between data analysis of physical and computer simulation experiments is statistical inference. In the above example, the F statistic is used to measure the impact of factors on the response variable. While statistical test and estimation can be used for analyzing data of physical experiments, they cannot be meaningfully applied to computer simulation experiments in terms of statistical inference because the results of computer experiments do not include any replication error.

This difference means that it is necessary to use a different interpretation of the analyzed results. According to experimental design theory, statistical inference plays an important role for identifying factor effects based on measurements that have replication error. However, it is important to consider the features of data collected by computer simulation in order to interpret the result precisely. Simply speaking, F and other statistics should be treated as descriptive statistics rather than inferences.

4 Optimization of response

4.1 Problem formulation

Background

After determining a set of important factors, it is necessary to optimize the response variables. This typically involves maximization and minimization of the response. Although it seems to be a simple problem from a mathematical viewpoint, various aspects need to be taken into consideration for practical optimization, such as the value of the response variable, the robustness of the results, ease of operation, cost and so forth.

In order to account for these various aspects, it is advantageous to use an approximate function to describe the relationship between response and factors. For example, an approximating function has robustness of response with respect to deviations of the factor levels from the assumed optimum values, while simple mathematical optimization does not include the concept of robustness and only gives the optimum value.

Furthermore, it is necessary to examine several different values of the factors in order to find a practically favorable combination of factors. Simple optimization, in general, gives only the optimum condition of the factors, therefore an approach that gives not only the optimized condition but also several sub-optimal, but favorable conditions is helpful. For this reason, approximate functions are helpful for investigating better conditions. For example once the relationship between the response and the factors is expressed as a polynomial function, the function gives the optimum value, robustness, and so forth.

The conventional approach for determining an approximating function is a combinatorial usage of composite design and fitting of second-order model by least squares. This approach was first developed by Box and Wilson (1951) and has been adapted and applied to many fields for a long time.

Formulation

At this stage, some active factors, say $(x_1, \ldots, x_q)$, has already have been identified using a computer simulation model $\mu(x_1, x_2, \ldots, x_q)$. The major aim is to find a function $\widehat{\mu}(x_1, x_2, \ldots, x_q)$ that is an approximation of the true response function $\mu(x_1, x_2, \ldots, x_q)$. As introduced in the previous section, the second order model:

$$\begin{aligned}\mu(x_1, x_2, \ldots, x_q) = \beta_0 &+ \beta_1 x_1 + \beta_2 x_2 + \cdots + \beta_q x_q \\ &+ \beta_{12} x_1 x_2 + \beta_{13} x_1 x_3 + \cdots + \beta_{q-1\,q} x_{q-1} x_q \\ &+ \beta_{11} x_1^2 + \beta_{22} x_2^2 + \cdots + \beta_{qq} x_q^2\end{aligned} \tag{1}$$

is widely applied.

In the above equation, the linear and quadratic terms denote the main effect of the factors, while the product terms denote the interaction effects of the two factors. For example, $\beta_1 x_1$ and $\beta_{11} x_1^2$ denote the main effect of factor x_1, while $\beta_{12} x_1 x_2$ denotes the interaction effect of the factors x_1 and x_2. Furthermore, the parameter used to estimate $\widehat{\beta}$ is derived by ordinary least squares.

In order to fit the second-order model shown in Equation (1), central composite design is widely applied in many fields, because it has many advantages including requiring a small number of experiments, a high estimate precision, the ability to detect higher-order effects, the ability to determine error variance, and so forth. More details are shown in the standard textbooks of response surface methodology, such as Box and Draper (1987) and Myers and Montgomery (2002).

While this approach is useful in many cases, there are some practical problems that cannot be solved by this approach. One typical example is the requirement to approximate a complicated function. The following section considers a complicated function associated with a practical model.

4.2 Example of wire bonding process — approximation of a complicated function —

Outline of the case

This subsection introduces an example of technology development using a computer simulation study for a wire-bonding process of an integrated circuit based on research by Masuda, Yamazaki and Yoshino (2005). The process produces an integrated circuit that requires more than one hundred wire bonds. The wire is connected on one side first, while the other side is bonded by using micro-joining at a certain frequency. The outline of the wire is shown in Figure 3.

Under some conditions, the connection point on the first side is broken because of resonance when the second side is being bonding. This resonance occurs under a certain combinations of wire conditions of x_1 (width), x_2 (diameter), and x_3 (height) and f (frequency) of the bonding machine. In order to solve this problem, a computer simulation model has been developed.

According to this model, the moment at the first side is calculated by a computer simulation when the values of x_1, x_2, x_3 and f are specified. An

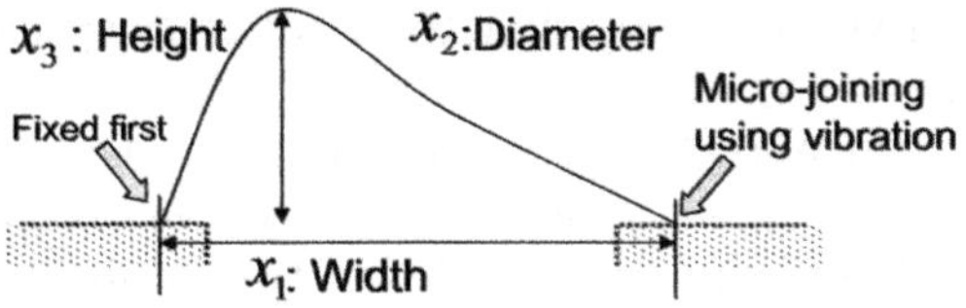

Fig. 3. Outline of the wire and its factors in this case study

example of computer outputs is shown in Figure 4, where the set of values are ($x_1 = 3.00\,(\text{mm})$, $x_2 = 0.03\,(\text{mm})$, $x_3 = 0.78\,(\text{mm})$) and ($x_1 = 3.14\,(\text{mm})$, $x_2 = 0.03\,(\text{mm})$, $x_3 = 1.28\,(\text{mm})$).

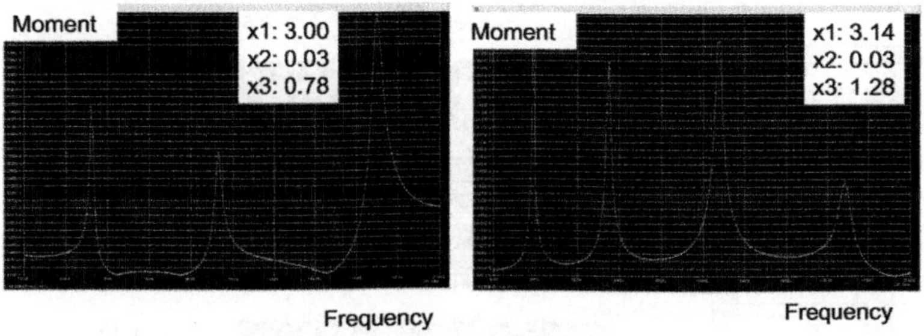

Fig. 4. Two examples of computer simulation output

The horizontal axis of this figure represents the frequency of the bonding machine and the vertical axis represents the moment at the first side. This figure implies that resonance occurs under the conditions represented by the point x_1, x_2, x_3 and f.

In this process, the values of x_1 (width) and x_2 (diameter) are specified by an a priori process, thus we need to find a good combination of x_3 (height) and f (frequency). Furthermore, the values of x_1 (width) varies in an integrated circuit. Therefore a robust condition for x_3 and f should be obtained for various values of x_1 and x_2.

Design to approximate a complicated function

This example is a different problem from determining an optimum condition. There are some requirements on the operating conditions, such as robustness of the conditions, cost. Therefore, an approximating function $M(x_1, x_2, x_3, f)$ to the moment force will be investigated as a function x_1, x_2, x_3 and f. Once a good approximating function has been found, better values of x_3 and f can be determined for given levels of x_1 and x_2. If the computation time was not so long, a brute-force computation might produce a good level of factors. However, it requires a long time to calculate the various combinations of $M(x_1, x_2, x_3, f)$.

In order to obtain an approximating function, we calculate the moment under some combination of x_1, x_2, x_3 along with various values of f. Specifically, x_1, x_2 is determined based on a uniform design that is able to fit various types of functions. Because of time constraints, the fifteen combinations of x_1 and x_3 are determined by uniform design that was obtained from the following web page `http://www.math.hkbu.edu.hk/~ktfang/`.

The factor x_2 (diameter) is treated as a qualitative factor because this process uses only either 0.03 or 0.04 (mm) diameter wires. The frequency is determined for 210 levels in order to cover the various ranges. An outline of the data is shown in Figure 5. Based on the collected data, an approximating function is can be found.

x1: witdh	x3: Height	x2 diameter
2.000	0.929	0.03
2.143	1.357	0.03
2.286	0.571	0.03
2.429	1.143	0.03
2.571	0.714	0.03
2.714	1.500	0.03
2.857	1.000	0.03
3.000	0.786	0.03
3.143	1.286	0.03
3.286	0.500	0.03
3.429	1.214	0.03
3.571	0.857	0.03
3.714	1.429	0.03
3.857	0.643	0.03
4.000	1.071	0.03
2.000	0.929	0.04
2.143	1.357	0.04
2.286	0.571	0.04
2.429	1.143	0.04
2.571	0.714	0.04
2.714	1.500	0.04
2.857	1.000	0.04
3.000	0.786	0.04
3.143	1.286	0.04
3.286	0.500	0.04
3.429	1.214	0.04
3.571	0.857	0.04
3.714	1.429	0.04
3.857	0.643	0.04
4.000	1.071	0.04

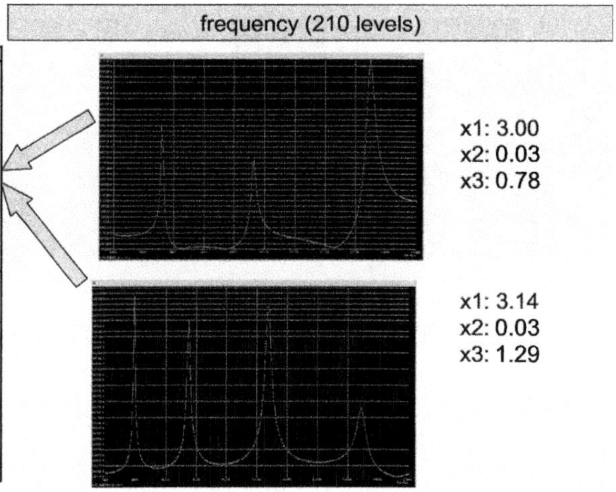

Fig. 5. Outline of the applied design and collected data

Data analysis to find an approximating function

Figure 4 shows a high level of complexity between the response and factors, which makes it difficult to fit a polynomial model to the data. We fit the model given by

$$M(x_1, x_2, x_3, f) = b_0 + b_1 f + \sum_{k=1}^{K} a_k \exp\left(\frac{f - m_k}{s_k}\right)^2 \tag{2}$$

in order to describe the relationship between the moments and the factors. The parameters b_0 and b_1 express a general trend, K denotes the number of peaks at the range of frequency.

A plot of Equation (2) is given in Figure 6. In the model, the exponential term describes the peak for each experimental run. This model is an application of the probability density function of normal distribution. In the field of neural network, this model is sometimes referred to as the Radial Basis Function that is utilized to describe the response to a stimulus in a neural network.

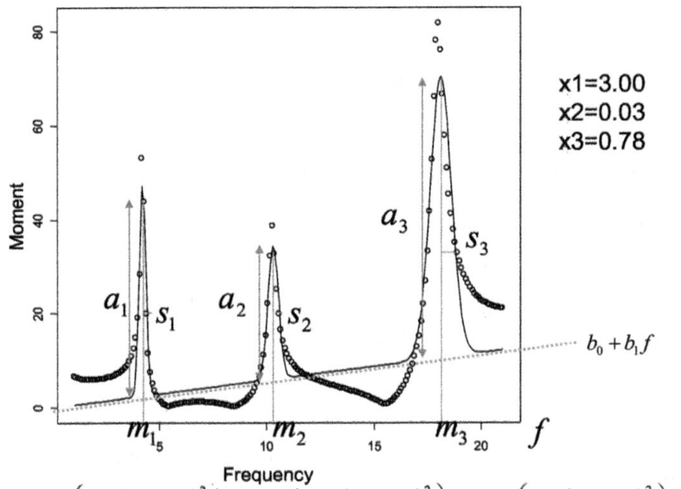

$$b_0 + b_1 f + a_1 \exp\left(-\frac{1}{2}\left(\frac{f-m_1}{s_1}\right)^2\right) + a_2 \exp\left(-\frac{1}{2}\left(\frac{f-m_2}{s_2}\right)^2\right) + a_3 \exp\left(-\frac{1}{2}\left(\frac{f-m_3}{s_3}\right)^2\right) + \ldots$$

Fig. 6. Approximation of the true response function

The first step involves estimating the parameters, b_0, b_1, a_k, m_k, s_k at each experimental run. Figure 6 also shows an example of fitting a model under the condition of $x_1 = 3.00$, $x_2 = 0.03$, $x_3 = 0.78$. Specifically, the value of K is 3 because three peaks can be found in the figure. The parameters are estimated by non-linear least squares using the data along with 210 levels of f. In a similar manner, the parameters in Equation (2) are estimated at each experimental runs.

The second step is to identify the relationship between the factors x_1, x_2 and x_3 and the parameters b_0, b_1, a_k, m_k, s_k. The estimates are derived by least squares where the parameters and factors are treated as response and input variables, respectively. The results are as follows:

$$\begin{aligned}
b_0 &= -96.1 + 15.6x_1 + 5640.0x_2 - 37.3x_3 + 19.1x_1x_2 - \\
b_1 &= 53.0 - 32.0x_1 - 1125.3x_2 + 16.6x_3 - 5.5x_1x_2 \\
&\quad + 433.7x_1x_3 + 4.1x_1^2 \\
a_1 &= -743.0 + 12.5x_1 + 32652.3x_2\, 252.9x_3 - 261.3x_1x_2
\end{aligned}$$

$$
\begin{aligned}
&\quad - 12456.0x_1x_3 + 22767.2x_2x_3 + 87.0x_1^2 \qquad (3)\\
m_1 &= 14.6 - 6.4x_1 + 391.5x_2 - 10.1x_3 + 2.8x_1x_2 - 59.3x_1x_3\\
&\quad - 86.5x_2x_3 + 0.6x_1^2 + 0.8x_2^2\\
s_1 &= 0.10 - 0.001x_1 + 0.1x_2 + 1.9x_3 - 0.1x_1x_2\\
&\vdots
\end{aligned}
$$

The above estimates are substituted into Equation (2) to obtain an approximation for the relation. In practice, the accuracy in estimating the parameters m_k is more crucial than the accuracy in estimating b_0, b_1, a_k and s_k because the frequency at which resonance occurs is more critical than the estimate of the moment in this problem. The coefficient of determination of the least squares indicates that the accuracy in estimating m is better than that for the other parameters. For example, the coefficient of determination of m_1 estimation is more than 90%. This fact implies that there is sufficient accuracy to estimate the resonance.

Figure 7 shows two estimated results based on the above approximation for various values of the factors. Specifically, two examples are calculated under (a) $(x_1 = 2.14, x_2 = 0.03, x_3 = 1.36)$ and (b) $(x_1 = 3.00, x_2 = 0.03, x_3 = 0.78)$ This figure suggests having enough accuracy for approximation as a whole. Specifically, the frequencies at which resonance occurs are estimated well in both figures. This fact is also confirmed in the other levels of x_1, x_2 and x_3. On the other hand, it seems to have a minor problem of estimating the moment under some levels of x_1, x_2 and x_3. For example, the approximating values of the moment seem to depart from the computer simulation values at the high frequency in Figure 7 (b). However, it is not a crucial problem because the primal focus of this example is to find the levels of x_1, x_2, x_3 and f at which resonance occur in order avoid breaking the bonded wire. In other words, the approximation of the moment level is not a primal focus.

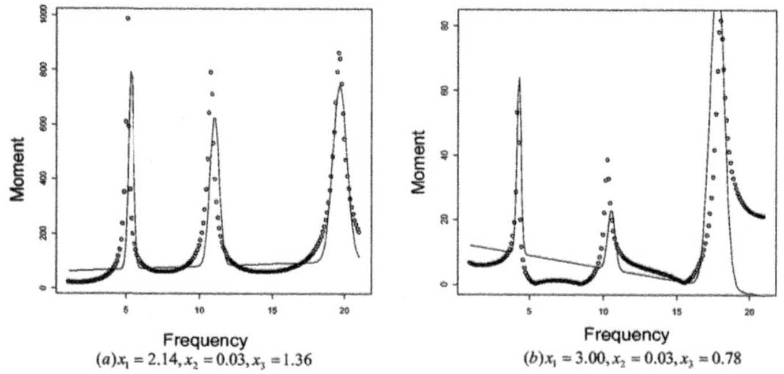

Fig. 7. Fitting the approximating function to the collected data

Application

In the process, an a priori process determines the width and the diameter based on product design and the height and frequency need to be determined by the process described above. In other words, the height and frequency have to be found once the levels of x_1 (width) and x_2 (diameter) have been determined. Furthermore, the values vary from product to product. The developed approximating function is helpful because it makes calculations by finite element software unnecessary.

An example of an application is shown in Figure 8. This figure is drawn using predetermined values for x_1 and x_2 and the marked region indicates the range of x_3 and f values that avoid resonance. In the past, several computer simulations had been had developed for the case when the levels of x_1 and x_2 were determined. However, this approximation shows a set of good conditions on x_3 and f without the need to perform a computer simulation study. The approximating function is helpful for initial product design.

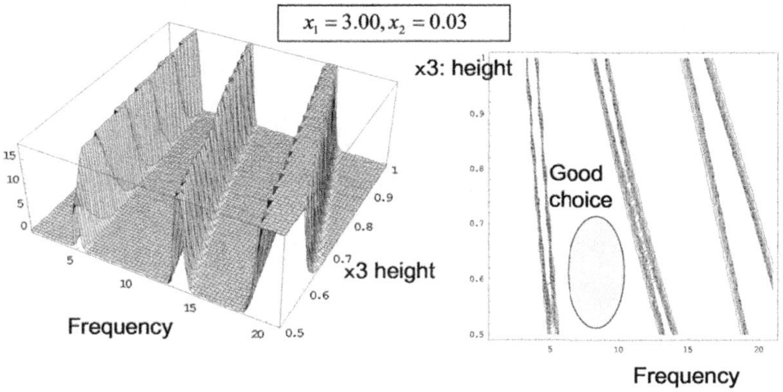

Fig. 8. An example of region of x_3 and f to avoid resonance under $x_1 = 3.00$ and $x_2 = 0.03$

5 Concluding remarks

Design of experiments has been applied in many practices of technology development when physical experiments are required to examine the facts. The three principles: *randomization*, *replication*, *blocking* developed by Fisher form the core of physical experiments. The importance of this method will continue in the future.

The recent advance in computer simulations requires advances in design of experiments. Examples of the advances are found in the books by Santner,

Williams and Notz, W. (2003) and Yamada (2004b). In case of computer simulation experiments, it is not required to consider replication under the same conditions, because the measured values do not include replication error. Thus the interpretation of the analyzed results needs to be modified. Specifically, the result should be considered as a descriptive statistical summary of the numerical data. Regarding the design technique for determining the optimum condition, there are various options for screening and approximating problems. Some good designs should be selected based on the characteristic features of the problem under consideration.

References

1. Booth, K.H.V. and Cox, D.R. (1962) Some systematic supersaturated designs. Technometrics, 4 (4) 489–495.
2. Box, G.E.P., and Wilson, K.B. (1951) On the experimental attainment of optimum Conditions. J. Royal Statist. Soc. Ser. B 13 1–45.
3. Box, G.E.P. and Draper, N.R. (1987) Empirical model-building and response surfaces, John Wiley & Sons, New York.
4. Fang, K.T., Lin, D.K.J., Winker, P. and Zhang, Y. (2000) Uniform design: theory and applications. Technometrics, 42, 237–248.
5. Fang, K.T., Lin D.K.J., Ma C.X. (2000) On the construction of multi-level supersaturated designs. J. Statist. Plann. Inference, 86 239–252.
6. Fisher, R.A. (1966) The design of experiments 8th ed., Oliver and Boyd, Edinburgh.
7. Lin, D.K.J. (1993) A new class of supersaturated designs. Technometrics, 35 (1) 28–31.
8. Lin, D.K.J. (1995) Generating systematic supersaturated designs. Technometrics, 37 (2) 213–225.
9. Myers, R.H. and Montgomery, D.C. (2002) Response Surface Methodology 2nd ed., John Wiley & Sons, New York.
10. Satterthwaite, F.E. (1959) Random balance experimentation (with discussion). Technometrics, 1 111–137.
11. Santner, T.J., Williams, B.J. and Notz, W. (2003) The Design and Analysis of Computer Experiments, Springer Series in Statistics, New York.
12. Taguchi, G. (1976) Design of experiments (1), Maruzen, Tokyo (in Japanese).
13. Wu, C.F.J. and Hamada, M. (2000) Experiments, John Wiley & Sons, New York.
14. Westfall, P.H., Young, S.S. and Lin, D.K.L. (1998) Forward selection error control in the analysis of supersaturated design. Statist. Sinica, 8 101–117.
15. Yamada, S., Ikebe, Y., Hashiguchi, H. and Niki, N. (1999) Construction of three-level supersaturated design. J. Statist. Plann. Inference, 81 183–193.
16. Yamada, S. and Lin, D.K.J. (1999) Three-level supersaturated design. Statist. Prob. Letters, 45 31–39.
17. Yamada, S. and Matsui, T. (2002) Optimality of mixed-level supersaturated designs. J. Statist. Plann. Inference, 104 459–469.
18. Yamada, S. (2004) Selection of active factors by stepwise regression in the data analysis of supersaturated design. Quality Engineering. 16, 501–513.

19. Yamada, S. (2004b) Design of Experiments — Methodology —, JUSE publications, Tokyo (in Japanese).
20. Yamazaki, Y., Masuda, M. and Yoshino, Y. (2005) Analysis of wire-loop resonance during AL wire bonding. Proc. 11th symposium on microjoining and assembly technology in electronics. February 3–5, 409–412.

Uniform Design in Computer and Physical Experiments*

Kai-Tai Fang[1] and Dennis K. J. Lin[2]

[1] Department of Mathematics, Hong Kong Baptist University, Kowloon Tong, Hong Kong, CHINA
[2] Department of Supply Chain and Information Systems, The Pennsylvania State University University Park, PA, USA

Summary. Computer experiments have been widely used in various fields of industry, system engineering, and others because many physical phenomena are difficult or even impossible to study by conventional experimental methods. Design and modeling of computer experiments have become a hot topic since late Seventies of the Twentieth Century. Almost in the same time two different approaches are proposed for design of computer experiments: Latin hypercube sampling (LHS) and uniform design (UD). The former is a stochastic approach and the latter is a deterministic one. A uniform design is a low-discrepancy set in the sense of the discrepancy, the latter is a measure of uniformity. The uniform design can be used for computer experiments and also for physical experiments when the underlying model is unknown. In this paper we review some developments of the uniform design in the past years. More precisely, review and discuss relationships of fractional factorial designs including orthogonal arrays, supersaturated designs and uniform designs. Some basic knowledge of the uniform design with a demonstration example will be given.

Key words: Computer Experiments, experimental design, factorial design, supersaturated design, uniform design

1 Motivation

Computer experiments and/or computer simulations have been widely used for studying physical phenomena in various fields of industry, system engineering, and others because many physical processes/phenomena are difficult or even impossible to study by conventional experimental methods. We describe the physical process by a mathematical model, implemented with code

* A keynote speech in *International Workshop on The Grammar of Technology Development*, January 17–18, 2005, Tokyo, Japan. The authors would thank the invitation.

on a computer. Design and modeling of computer experiments have become a hot topic since late Seventies of the Twentieth Century. Motivated by three big projects in system engineering in 1978 Prof. Y. Wang and myself in [4, 36] proposed the so-called the uniform experimental design or uniform design (UD) for short. There were six or more input variables in one of the three projects and the output y can be obtained by solving a system of differential equations. It costed one day calculation from an input to the corresponding output. Clearly, the relationship between the input and output has no analytic formula and is very complicated. The true model can be expressed as

$$y = f(x_1, \ldots, x_s) \equiv f(\boldsymbol{x}), \qquad \boldsymbol{x} \in T, \tag{1.1}$$

where $\boldsymbol{x} = (x_1, \ldots, x_s)$ is the input, y the output, T the experimental domain and function f is known and has no analytic expression. The engineers wanted to find a simple and approximate model or called as a metamodel

$$y = g(x_1, \ldots, x_s) = g(\boldsymbol{x}) \tag{1.2}$$

such that the difference of $|f(\boldsymbol{x}) - g(\boldsymbol{x})|$ is small over the domain T in a certain sense. The metamodel g should be much easy to compute, i.e., the computation complexity for $g(\boldsymbol{x})$ is much less than one for $f(\boldsymbol{x})$. For searching a good metamodel it is suggested to choose a set of points, $\boldsymbol{x}_1, \ldots, \boldsymbol{x}_n$ in T and calculate their corresponding outputs to form a data set $\{(\boldsymbol{x}_i, y_i),\ i = 1, \ldots, n\}$. Then applying some useful modeling techniques to find a good model to fit the data. If the chosen model can predict the output at any point in T well, this model can be regarded as a metamodel. This created a new concept: *design and modeling for computer experiments* (DMCE) (or DACE, design and analysis for computer experiments) in that era. Figure 1 shows the idea of computer experiments.

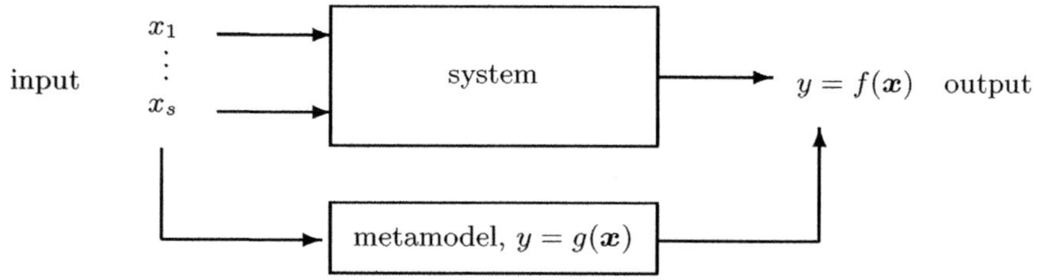

Fig. 1. Computer Experiments

We may have several goals in computer experiments, such as a) to explore the relationships between the input and the output; b) To find maximum/minimum value of y and the corresponding input-combination; c) to quantify the leverage of each input variable to the output; d) to determinate variables for which the values are optimized to minimize the variability of response variable, y.

Almost in the same time the author met many industrial experiments, the related engineer did not know the underlying model well. We can express the model for these experiments by

$$y = f(x_1, \ldots, x_s) + \varepsilon \equiv f(\boldsymbol{x}) + \varepsilon, \qquad \boldsymbol{x} \in T, \tag{1.3}$$

where $f(\boldsymbol{x})$ is unknown, maybe nonlinear, the domain T may be large, and ε is the random error. The experimenter wants to find a model

$$y = g(\boldsymbol{x}) + \varepsilon \tag{1.4}$$

for estimation of the true model by an experimental data such that $g(\boldsymbol{x})$ is very close to $f(\boldsymbol{x})$ in a certain sense. Both of computer experiments and physical experiments with model unknown meet the same aspects:

A. Experimental Design: Note that the true model in many studies may have the following complexities: 1) multi-factor: the number of factors may be high; 2) non-linearity: the function $f(\boldsymbol{x})$ is a non-linear function in $\boldsymbol{x}$; and 3) large domain: the experimental domain is large so that $f(\boldsymbol{x})$ may have more than one local minimum/maximum point. Due to the above complexities we want to choose experimental points uniformly scattered in the domain so that we can have a good estimation for the true model. This leads to the terminology "uniform design", or "space-filling design".

B. Modeling: We wish to find a high quality metamodel that approximates the true model well over the domain. A good metamodel should have less computation complexity and easy to explore relationship between the input variables and the output.

Computer experiments have no random errors while physical experiments have measurement errors, environment effects and so on. This fact leads to much different between computer experiments and physical experiments with model unknown. For example, three well-known principles for physical experiments: *replication, randomization* and *blocking* are meaningless for computer experiments. Various statistical methods can't be directly used for analyzing data from computer experiments. However, the idea of many statistical methods can still be applied for modeling computer experiments. We shall discuss this issue late.

There are many approaches to computer experiments. In this paper I focus only on the uniform design for its theory, method and recent development. In Section 2 an introduction to theory and methodology of the uniform design is given, and a demonstration example shows implementation of the uniform design to physical experiments with model unknown. Section 3 introduces some recent development of the uniform design. Some applications of the uniform design will be given in Section 4.

2 Theory and Methodology of The Uniform Design

2.1 Theory

Suppose that the purpose of experimental design is to find a metamodel $g(\boldsymbol{x})$ in (1.2) to approximate the true model $f(\boldsymbol{x})$ in (1.1). How do we measure the closeness of $g(\boldsymbol{x})$ to $f(\boldsymbol{x})$? There are many criteria. A natural idea in engineering requests the metamodel $g(\boldsymbol{x})$ satisfying

$$\text{diff}(f, g) = |f(\boldsymbol{x}) - g(\boldsymbol{x})| < \delta, \qquad \boldsymbol{x} \in T, \tag{2.1}$$

where δ is the given accuracy in the project. For simplicity, without loss of any generality, the experimental region can be assumed to be a unit cube $C^s = [0, 1]^s$. Another criterion is

$$\text{p-diff}(f, g) = \int_{C^s} |f(\boldsymbol{x}) - g(\boldsymbol{x})|^p \, d\boldsymbol{x} < \delta, \tag{2.2}$$

where $p > 0$, more often choosing $p = 1$ or $p = 2$. The class of functions $f(\boldsymbol{x})$ can be considered the class of L_2-integrable continuous functions, denoted by $\mathcal{G}_2$. It is not easy to find a unified way to construct metamodel $g(\boldsymbol{x})$ satisfying the criterion (2.1) or (2.2) for each function in $\mathcal{G}_2$. Therefore, the so-called *overall mean model* was suggested. Let $\mathcal{P} = \{\boldsymbol{x}_1, \ldots, \boldsymbol{x}_n\}$ is a set of experimental points on C^s. Suppose the experimenter wants to estimate the overall mean

$$\text{mean}(y \,|\, C^s) = \int_{C^s} f(\boldsymbol{x}) \, d\boldsymbol{x}$$

by the sample mean

$$\bar{y}(\mathcal{P}) = \frac{1}{n} \sum_{i=1}^{n} f(\boldsymbol{x}_i). \tag{2.3}$$

We wish to find a design $\mathcal{P}$ such that diff-mean $\equiv |\text{mean}(y \,|\, C^s) - \bar{y}(\mathcal{P})|$ as small as possible. The famous Koksma-Hlawka inequality in quasi-Monte Carlo methods provides a upper bound of the difference

$$\text{diff-mean} = |\text{mean}(y \,|\, C^s) - \bar{y}(\mathcal{P})| \leq V(f) D(\mathcal{P}), \tag{2.4}$$

where $D(\mathcal{P})$ is the star discrepancy of $\mathcal{P}$, a measure of uniformity of the set $\mathcal{P}$ over the domain C^s and does not depend on f, and $V(f)$ is the total variation of the function f in the sense of Hardy and Krause (see Niederreiter [30]).

Let $F_{\mathcal{P}}(\boldsymbol{x})$ be the empirical distribution function

$$F_{\mathcal{P}}(\boldsymbol{x}) = \frac{1}{n} \sum_{i=1}^{n} I(\boldsymbol{x} \,|\, \boldsymbol{x}_i \leq \boldsymbol{x}),$$

where $\boldsymbol{x}_i \leq \boldsymbol{x}$ means each component of $\boldsymbol{x}_i$ is less than or equals to the corresponding component of $\boldsymbol{x}$, $I(\boldsymbol{x} \,|\, \boldsymbol{x}_i \leq \boldsymbol{x})$ is the indicate function such that it equals to 1 if $\boldsymbol{x}_i \leq \boldsymbol{x}$, otherwise zero. The star discrepancy is defined as

$$D(\mathcal{P}) = \max_{\boldsymbol{x} \in C^s} |F_{\mathcal{P}}(\boldsymbol{x}) - F(\boldsymbol{x})|, \tag{2.5}$$

where $F(\boldsymbol{x})$ is the uniform distribution function on C^s.

The Koksma-Hlawka inequality indicates:

a) The lower the star discrepancy, the better uniformity the set of points has. This suggests to minimize the star discrepancy $D(\mathcal{P})$ on all designs of n runs on C^s, i.e., to find a *uniform design*. Fang and Wang (Fang [4] and Wang and Fang [36]) proposed the uniform design and provides a number of uniform designs. When the number of runs, n increases, roughly speaking, one can find design $\mathcal{P}_n$ such that $D(\mathcal{P}_n)$ decreases.

b) The uniform design is **robust** against the model specification. For example, two models $y = f_1(\boldsymbol{x})$ and $y = f_2(\boldsymbol{x})$ have the same variation $V(f_1) = V(f_2)$, a uniform design may have the same level performance for these two models.

c) If the true model $f(\boldsymbol{x})$ has a large variation, in general, we need more runs to reach the same upper bound of diff-mean.

d) There are many versions of the Koksma-Hlawka inequality, where the star discrepancy $D(\mathcal{P})$ is replaced by another discrepancy and the total variation $V(g)$ is defined according to the definition of the given discrepancy. Hickernell [20] gave a comprehensive discussion and proposed some new measures of uniformity, among of which the centered L_2-discrepancy (CD) and the wrap-around L_2-discrepancy (WD) have good properties and satisfies the Koksma-Hlawka inequality. The CD and WD have nice computational formulas

$$(CD(\mathcal{P}))^2 = \left(\frac{13}{12}\right)^s - \frac{2}{n}\sum_{k=1}^{n}\prod_{j=1}^{s}\left(1 + \frac{1}{2}|x_{kj} - 0.5| - \frac{1}{2}|x_{kj} - 0.5|^2\right)$$
$$+ \frac{1}{n^2}\sum_{k=1}^{n}\sum_{j=1}^{n}\prod_{i=1}^{s}\left[1 + \frac{1}{2}|x_{ki} - 0.5| + \frac{1}{2}|x_{ji} - 0.5| - \frac{1}{2}|x_{ki} - x_{ji}|\right], \tag{2.6}$$

and

$$(WD(\mathcal{P}))^2 = \left(\frac{4}{3}\right)^s + \frac{1}{n^2}\sum_{k=1}^{n}\sum_{j=1}^{n}\prod_{i=1}^{s}\left[\frac{3}{2} - |x_{ki} - x_{ji}|(1 - |x_{ki} - x_{ji}|)\right], \tag{2.7}$$

respectively, where $\boldsymbol{x}_k = (\boldsymbol{x}_{k1}, \ldots, x_{ks})$ is the kth experimental point.

Obviously, the overall mean model is too simple and may not reach the task: estimation of the true model $f(\boldsymbol{x})$. But, the overall mean model provides a simple way to develop methodology and theory of the uniform design. It is surprising that the uniform design has an excellent performance for both computer experiments and physical experiments with model unknown.

Wiens [37] concerned with designs for approximately linear regression models and show that the uniform design measure (the uniform distribution on C^s) is maximin in the sense of maximizing the minimum bias in the regression estimate of σ^2 and is also minimax in the sense of minimizing the maximum bias in the regression estimate of σ^2. Xie and Fang [41] pointed out that the uniform measure is admissible and minimax under the model

$$y = f(x_1, \ldots, x_s) + \epsilon,$$

where f is unknown, but belongs to some function family. Hickernell [21] considered robust regression models

$$y = f(\boldsymbol{x}) + \varepsilon = \mathrm{mean}(y) + h(\boldsymbol{x}) + \varepsilon,$$

where the function $f(\boldsymbol{x})$ is decomposed into the overall mean value of $f(\boldsymbol{x})$ and mis-specification $h(\boldsymbol{x})$. He proposed two models *average mean-square-error model* and *maximum mean-square-error model*. With a certain condition he proved that the uniform design is optimal under these models. His results show that the uniform design is robust for model specification. Hickernell and Liu [22] consider efficiency and robustness of experimental design. They said "Although it is rare for a single design to be both maximally efficient and robust, it is shown here that uniform designs limit the effects of aliasing to yield reasonable efficiency and robustness together."

2.2 Methodology

In this subsection we introduce how to apply the uniform design to real experiments. A uniform design for an experiment of s factors with n runs on the domain C^s is a set of n points such that this set has the minimum discrepancy, the latter can be centered L_2-discrepancy or others. If the domain T is a rectangle in R^s, a linear transformation can transfer n points on C^s into T.

When there is only one factor on the range $[a, b]$ in the experiment with n runs. The uniform design arranges n runs as $\{a+(b-a)\frac{1}{2n}, a+(b-a)\frac{3}{2n}, \ldots, a + (b-a)\frac{2n-1}{2n}\}$.

For multi-factor experiments it is not tractable to find a set of n points, $\mathcal{P} = \{\boldsymbol{x}_1, \ldots, \boldsymbol{x}_n\} \subset C^s$, such that it has the minimum discrepancy. Therefore, many authors focus on lattice points and introduce the concept of U-type designs.

Definition 1. A *U-type design* denoted by $U(n; q_1 \times \cdots \times q_s)$ is an $n \times s$ matrix with q_j entries at the jth columns such that the q_j entries appear in this column equally often. When some q_j are equal, we denote it by $U(n; q_1^{r_1} \times \cdots \times q_m^{r_m})$ with $r_1 + \cdots + r_m = s$. When all the number of q_j are equal to q, we write $U(n; q^s)$ and the corresponding designs are called symmetric, otherwise asymmetric or U-type design with mixed levels. Let $\mathcal{U}(n; q_1 \times \cdots \times q_s)$ be the set of all U-type designs $U(n; q_1 \times \cdots \times q_s)$. Similarly we have notations $\mathcal{U}(n; q_1^{r_1} \times \cdots \times q_m^{r_m})$ and $\mathcal{U}(n; q^s)$.

Very often we choose q entries in one column as $\{1, 2, \ldots, q\}$. Sometimes, q entries are chosen as $\{\frac{1}{2q}, \frac{3}{2q}, \ldots, \frac{2q-1}{2q}\}$. Let $\boldsymbol{U} = (u_{ij})$ be a U-type design in $\mathcal{U}(n; q_1 \times \cdots \times q_s)$ with entries $\{1, \ldots, q_j\}$ at the jth column. Take the transformation

$$x_{ij} = \frac{u_{ij} - 0.5}{q_j}, \qquad i = 1, \ldots, n, \ j = 1, \ldots, s. \tag{2.8}$$

and denote $\boldsymbol{X}_u = (x_{ij})$. Then $\boldsymbol{X}_u$ is a U-type design with entries $\{\frac{1}{2q_j}, \frac{3}{2q_j}, \ldots, \frac{2q_j-1}{2q_j}\}$ at the jth column. The matrix $\boldsymbol{X}_u$ is called the *induced matrix* of $\boldsymbol{U}$. The n rows of the matrix $\boldsymbol{X}_u$ are n points on $[0,1]^s$. Most measures of uniformity are defined on $[0,1]^s$ in the literature. Therefore, we define uniformity of a U-type design $\boldsymbol{U}$ through its induced matrix by

$$D(\boldsymbol{U}) = D(\boldsymbol{X}_u). \tag{2.9}$$

Definition 2. A design $\boldsymbol{U} \in \mathcal{U}(n; q_1 \times \cdots \times q_s)$ is called a uniform design under the pre-decided discrepancy D if

$$D(\boldsymbol{U}) = \min_{\boldsymbol{V} \in \mathcal{U}(n; q_1 \times \cdots \times q_s)} D(\boldsymbol{V}),$$

and is denoted by $U_n(q_1 \times \cdots \times q_s)$.

Under the CD in (2.6) Table 1 and Table 2 give two uniform designs $U_{12}(12^4)$ and $U_6(3^2 \times 2)$, respectively. Table 1 can arrange an experiment having at most 4 12-level factors with 12 runs and Table 2 can apply to an experiment of 6 runs and 3 factors where two have three levels and one has two levels. Figure 2 gives scatter plots for any two columns of $U_{12}(12^4)$. By a visualization we can see that 12 points on any two marginal square are uniformly scattered. How to construct uniform design tables is a challenging job. A comprehensive review on construction of uniform designs can refer to Fang and Lin [11] and Fang, Li and Sudjianto [10]. A number of UD tables can be found on the UD web site at `http://www.math.hkbu.edu.hk/UniformDesign`.

Table 1. $U_{12}(12^4)$

No	**1**	**2**	**3**	**4**
1	1	10	4	7
2	2	5	11	3
3	3	1	7	9
4	4	6	1	5
5	5	11	10	11
6	6	9	8	1
7	7	4	5	12
8	8	2	3	2
9	9	7	12	8
10	10	12	6	4
11	11	8	2	10
12	12	3	9	6

Table 2. $U_6(3^2 \times 2)$

No	**1**	**2**	**3**
1	1	1	1
2	2	1	2
3	3	2	1
4	1	2	2
5	2	3	1
6	3	3	2

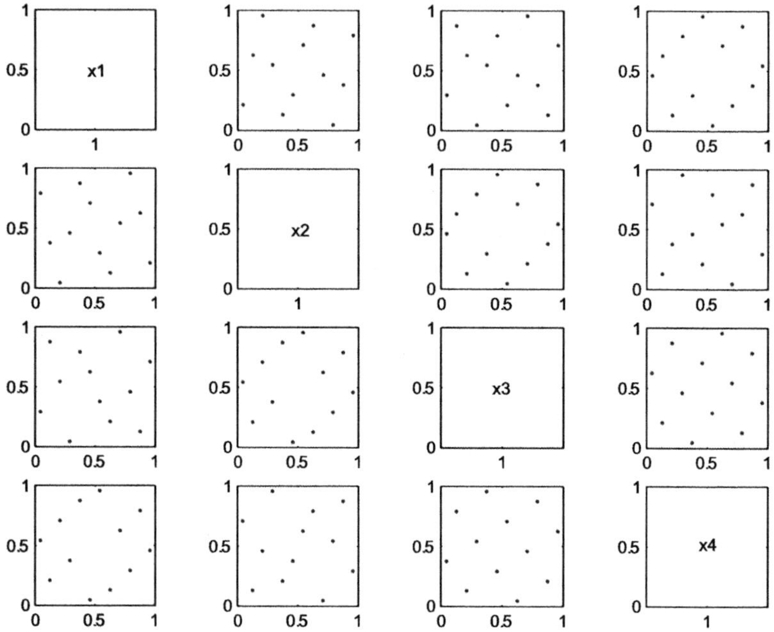

Fig. 2. Scatter plots of any two columns for $U_{12}(12^4)$

2.3 Demonstration example

For illustration applications of the uniform design consider a chemical experiment that is conducted in order to find the best setup to increase the yield. Four factors and 12 levels for each factor are chosen as follows:

x_1, the amount of formaldehyde (mol/mol): 1.0, 1.4, 1.8, 2.2, 2.6, 3.0, 3.4, 3.8, 4.2, 4.6, 5.0, 5.4
x_2, the reaction temperature (hour): 5, 10, 15, 20, 25, 30, 35, 40, 45, 50, 55, 60
x_3, the reaction time (hour): 1.0, 1.5, 2.0, 2.5, 3.0, 3.5, 4.0, 4.5, 5.0, 5.5, 6.0, 6.5
x_4, the amount of potassium (ml): 15, 20, 25, 30, 35, 40, 45, 50, 55, 60, 65, 70

The response variable is designated as the yield (y). This experiment could be arranged with a UD table of the form $U_n(12^4)$, where 12 is a divisor of n. It turns out that the experimenter chooses $U_{12}(12^4)$ design. The 12 levels marked by $1, 2, \ldots, 12$ are transformed into the real levels of the factors. It results in a design listed in Table 3 where the values in the bracket in the columns x_1, x_2, x_3 and x_4 are from the table $U_{12}(12^4)$. Run 12 experiments according to these 12 level-combinations and record the corresponding yield y (see the last column of Table 3).

The experimenters wanted to find a good metamodel such that they could find a level-combination of the factors with a yield that is much higher than

Table 3. Design and response

No	x_1	x_2	x_3	x_4	y
1	1.0 (1)	50 (10)	2.5 (4)	45 (7)	0.0795
2	1.4 (2)	25 (5)	6.0 (11)	25 (3)	0.0118
3	1.8 (3)	5 (1)	4.0 (7)	55 (9)	0.0109
4	2.2 (4)	30 (6)	1.0 (1)	35 (5)	0.0991
5	2.6 (5)	55 (11)	5.5 (10)	65 (11)	0.1266
6	3.0 (6)	45 (9)	4.5 (8)	15 (1)	0.0717
7	3.4 (7)	20 (4)	3.0 (5)	70 (12)	0.1319
8	3.8 (8)	10 (2)	2.0 (3)	20 (2)	0.0900
9	4.2 (9)	35 (7)	6.5 (12)	50 (8)	0.1739
10	4.6 (10)	60 (12)	3.5 (6)	30 (4)	0.1176
11	5.0 (11)	40 (8)	1.5 (2)	60 (10)	0.1836*
12	5.4 (12)	15 (3)	5.0 (9)	40 (6)	0.1424

the current one. We have mentioned that there are many ways to construct a metamodel. In this example, they considered only linear and quadratic regression models.

Note that the task for the experimenter is to find the best level-combination of the factors that can maximize the yield. The best result among the 12 responses is $y_{11} = 18.36\%$ at $x_1 = 5.0$, $x_2 = 40$, $x_3 = 1.5$ and $x_4 = 60$. This can be served as a benchmark. Is there any level-combination to produce a larger amount of yield? The simplest approximate model is the first-order regression or its submodel:

$$E(y) = \beta_0 + \beta_1 x_1 + \beta_2 x_2 + \beta_3 x_3 + \beta_4 x_4.$$

Unfortunately, we can't find a satisfactory result under this model as well as its submodel. Therefore, a more flexible second-order regression is considered as below

$$E(y) = \beta_0 + \sum_{i=1}^{4} \beta_i x_i + \sum_{i \le j} \beta_{ij} x_i x_j. \tag{2.10}$$

With a technique of model selection, we find a good submodel:

$$\begin{aligned} \hat{y} = {} & 0.0446 + 0.0029 x_2 - 0.0260 x_3 + 0.0071 x_1 x_3 \\ & + 0.000036 x_2 x_4 - 0.000054 x_2^2 \end{aligned} \tag{2.11}$$

with $R^2 = 97.43\%$ and $s^2 = 0.0001$. In the literature, the centered quadratic regression model

$$E(y) = \beta_0 + \sum_{i=1}^{4} \beta_i (x_i - \bar{x}_i) + \sum_{i \le j} \beta_{ij} (x_i - \bar{x}_i)(x_j - \bar{x}_j), \tag{2.12}$$

is also recommended, where $\bar{x}_i$ is the sample mean of x_i. In this data set, $\bar{x}_1 = 3.2$, $\bar{x}_2 = 32.5$, $\bar{x}_3 = 3.75$ and $\bar{x}_4 = 42.5$. Once again, by using some model selection technique, a metamodel is

$$\hat{y} = 0.1277 + 0.0281(x_1 - 3.2) + 0.000937(x_2 - 32.5) + 0.00114(x_4 - 42.5) \\ + 0.00058(x_3 - 3.75)(x_4 - 42.5) - 0.000082(x_2 - 32.5)^2 \quad (2.13)$$

with $R^2 = 97.05\%$ and $s^2 = 0.0002$.

By an ANOVA analysis and statistical diagnostics both models (2.11) and (2.13) are acceptable. Each can give an explanation for relationships among the four factors and the yield, but these relationships may be consistent or may be not consistent as the number of runs is small for such a four factor experiment. The experimenter should be carefully to make his/her own conclusion. The models (2.11) and (2.13) can be used to predict response at any point of the experimental domain. It also can be used for searching the 'best' combination of the factor-value. We maximize y with respect to x_i, $i = 1, \ldots, 4$ under models (2.11) or (2.13), respectively, over the domain, $\mathcal{X}$, that is to find x_i^*, $i = 1, \ldots, 4$ such that

$$\hat{y}(x_1^*, x_2^*, x_3^*, x_4^*) = \max_{\mathcal{X}} \hat{y}(x_1, x_2, x_3, x_4),$$

where $\hat{y}(x_1, x_2, x_3, x_4)$ is given by (2.11) or (2.13) respectively. By some optimization algorithm, it is easily found that under model (2.11), $x_1^* = 5.4$, $x_2^* = 50.2$, $x_3^* = 1$, $x_4^* = 70$ and the corresponding response $\hat{y}(5.4, 50.2, 1, 70) = 19.3\%$ is the maximum; and under model (2.13), $x_1^* = 5.4$, $x_2^* = 43.9$, $x_3^* = 6.5$, $x_4^* = 70$ and the corresponding response $\hat{y}(5.4, 43.9, 6.5, 70) = 26.5\%$ is the maximum. As two optimal points $\boldsymbol{x}_1^* = (5.4, 50.2, 1, 70)$ and $\boldsymbol{x}_2^* = (5.4, 43.9, 6.5, 70)$ do not appear in the plan (Table 3) some additional experiments are necessary for judging which metamodel is closed to the real one.

A simplest way is to implement m runs at these two optimal points $\boldsymbol{x}_1^*$ and $\boldsymbol{x}_2^*$ and to compare their mean yield. In this experiment the experimenter implemented three runs at $\boldsymbol{x}_1^*$ and $\boldsymbol{x}_2^*$ and find that the mean of y is 20.1% at $\boldsymbol{x}_1^*$ and 26.3% at $\boldsymbol{x}_2^*$, respectively. Thus we prefer (2.13), the centered quadratic regression model, as our chosen metamodel. Note that both metamodels recommend $x_1 = 5.4$ and $x_4 = 70$. This fact implies that we should consider increase upper bound of the experimental level for x_1 and x_4. The experimenter should consider a further investigate and arrange a consequent experiment.

2.4 Modeling techniques

We have considered the quadratic regression model and the centered quadratic regression model for the above example. It indicates that for an experiment we may have many possible metamodels. Many metamodels can be represented as a linear combination of a set of basis functions: $\{B_1(\boldsymbol{x}), B_2(\boldsymbol{x}), \ldots, B_m(\boldsymbol{x})\}$ defined on the experimental domain. In this case a metamodel g has of the form of $\sum_{j=1}^{m} \beta_j B_j(\boldsymbol{x})$ and

$$y(\boldsymbol{x}) = \sum_{j=1}^{m} \beta_j B_j(\boldsymbol{x}) + \varepsilon(\boldsymbol{x}), \tag{2.14}$$

where $\boldsymbol{x}$ is a point in the domain, β_j's are unknown coefficients to be estimated and $\varepsilon(\boldsymbol{x})$ is the random error. In fact, simply linear model, and both quadratic regression model and centered quadratic regression model are special cases of (2.14). For a univariate x-variable, the power spline basis has the following general form of $\sum_{j=1}^{m} \beta_j B_j(\boldsymbol{x})$ and

$$1,\ x,\ x^2,\ \ldots,\ x^p,\ (x-\kappa_1)_+^p,\ \ldots,\ (x-\kappa_K)_+^p, \tag{2.15}$$

where $\kappa_1, \ldots, \kappa_K$ are a set of selected knots, and a_+ stands for the positive part of a, i.e., $a_+ = aI(a > 0)$. Multivariate spline basis may be constructed from the univariate spline basis using the tensor product approach. The number of basis functions is often large, various techniques of variable selection are very useful for choosing a good submodel of (2.14) as a metamodel.

The Kriging model

$$y(\boldsymbol{x}) = \sum_{j=1}^{m} \beta_j B_j(\boldsymbol{x}) + z(\boldsymbol{x}), \qquad \boldsymbol{x} \in T, \tag{2.16}$$

has been widely used in modeling computer experiments, where $B_j(\boldsymbol{x})$ are given and β_j are unknown parameters; $z(\boldsymbol{x})$ is a stochastic fields, mostly choosing a stationary Gaussian field that has a zero mean function and a given covariance structure with some unknown parameters to be estimated. As we know that Kriging predictor interpolates its training data, this property is much suitable for modeling data from computer experiments where there is no random error, but it is not so good for the data from physical experiments. Therefore, the so-called "empirical Kriging model" is suggested

$$y(\boldsymbol{x}) = \sum_{j=1}^{m} \beta_j B_j(\boldsymbol{x}) + z(\boldsymbol{x}) + \varepsilon(\boldsymbol{x}), \tag{2.17}$$

where $\varepsilon(\boldsymbol{x})$ is a random error and is assumed to be uncorrelated with $z(\boldsymbol{x})$. A comprehensive study on Kriging model and empirical Kriging model can refer to Sacks et al. [32], Santner et al. [33] and Stein [35]. Other modeling techniques involve *neural networks, radial basis function model, local polynomial regression* and *Bayesian approach.* The reader can refer to Fang et al. [10] for the details.

The following points in modeling should be emphasized:

- There are many possible metamodels for an experiment. One should consider the following aspects and then choose one for the further process: 1) the metamodel gives a good prediction, 2) the response estimator can be easily calculated by the metamodel, 3) the metamodel can be easily explore relationships between the input factors and the output.

- Both computer experiments and physical experiments can share many common modeling techniques.
- The metamodel generated by some methods such as linear regression, quadratic regression, and of the form (2.14) can be easily interpreted, but the metamodel obtained by the neural networks, Kriging model and empirical Kriging model is not easily to directly give a clear exploration. For solving this difficulty the so-called *sensitivity analysis* (SA) has been developed. It studies how the variation in the output of a model can be apportioned, quantitatively, to different sources of variation and how the given model depends upon the information fed into it. The SA is used to provide an understanding of how the model response variables respond to changes in the inputs. A comprehensive study on SA can refer to Saltelli et al. [34].

3 Some Recent Development of The Uniform Design

The uniform design was based on quasi-Monte Carlo methods and its original theoretic proofs were mostly based on the number theory, not on statistics. There are many essential difficulties in development of its own theory:

- The uniformity is a geometrical criterion, it needs some justification in statistical sense;
- Initially, the uniform design theory is based on the quasi-Monte Carlo methods. The useful tool is the number theory. Most statisticians are lack of knowledge of the number theory;
- The overall mean model is far from the request of modeling;
- Construction of uniform design is a NP hard problem. It needs some powerful algorithms in optimization.

There was a rapid development in theory, methodology and applications of the uniform design in the past years, especially in the past ten years. It needs a very large space to review all the new results. Therefore, I mainly focus on relationship among fractional factorial design, supersaturated design and uniform design in this section.

3.1 Fractional factorial designs and supersaturated designs

Let us review some basic knowledge on these designs.

Definition 3. *For an experiment of n runs, s factors each having $q_1, \ldots, q_s$ levels respectively. A factorial design is a set of n level-combinations. A design where all the level-combinations of the factors appear equally often is called a full factorial design or a full design.*

The number of runs in a full factorial design should be $n = k\prod_{j=1}^{s} q_j$, where q_j is the number of levels of the factor j and k is the number replications for all the level-combinations. When all the factors have the same number of levels, q say, $n = kq^s$. In this case the number of runs of a full factorial design increases exponentially as the number of factors increases. Therefore, we consider to implement a subset of all the level-combinations that have a good representation of the complete combinations, this subset is called fractional factorial design (FFD for short). The most important and popularly used FFD is the orthogonal array.

Definition 4. *An orthogonal array (OA) of strength r with n runs and s factors each having q levels, denoted by $OA(n, s, q, r)$, is a FFD where any subdesign of n runs and r factors is a full design. When $r = 2$, the notation $L_n(q^s)$ is often for $OA(n, s, q, 2)$ in the literature.*

Strength two orthogonal arrays are extensively used for planning experiments in various fields and are often expressed as orthogonal design tables. Table 4 presents two $L_9(3^4)$, where left one $L_9(3^4)_1$ can be found in most textbook while the right one $L_9(3^4)_2$ was obtained by Fang and Winker [17]. Both can arrange an experiment of nine runs and at most four factors each having 3 levels. Is is easy to check that these two designs are isomorphic (see section 3.4 for the definition of the isomorphism). From the traditional view these two designs are equivalent. However, it is easy to find that $L_9(3^4)_2$ has a smaller CD-value than $L_9(3^4)_1$ has. Fang and Ma (2000) found some differences in statistical inference between the two designs. This gives an important message that uniformity of the design can provide additional information in statistical ability of the design. The reader can refer Dey and Mukerjee [3] and Hedayat, Sloane and Stufken [19] for the details.

Table 4. Two $L_9(3^4)$ Tables

No	$L_9(3^4)_1$				$L_9(3^4)_2$			
1	1	1	1	1	1	1	1	2
2	1	2	2	2	1	2	3	1
3	1	3	3	3	1	3	2	3
4	2	1	2	3	2	1	3	3
5	2	2	3	1	2	2	2	2
6	2	3	1	2	2	3	1	1
7	3	1	3	2	3	1	2	1
8	3	2	1	3	3	2	1	3
9	3	3	2	1	3	3	3	2

The number of runs for orthogonal array $OA(n, s, q, 2)$ is at least q^2. In industrial and scientific experiments, especially in their preliminary stages, very often there are a large number of factors to be studied and the run size is limited because of expensive costs. However, in many situations only a few factors are believed to have significant effects. Under the effect sparsity assumption, supersaturated designs have been suggested and can be effectively used to identify the dominant factors. The reader can refer to Yamada and Lin [43] and Lin [25] for a comprehensive introduction and recent development.

Definition 5. *Supersaturated designs are fractional factorials in which the number of estimated (main or interaction) effects is greater than the number of runs. Consider a design of* n *runs and* s *factors each having* q *levels. The design is called unsaturated if* $n-1 > s(q-1)$; *saturated if* $n-1 = s(q-1)$; *and supersaturated if* $n-1 < s(q-1)$.

Note that there are some common aspects among the orthogonal array, supersaturated design and uniform design:

- they are subset of level-combinations of the factors
- they are constructed based on U-type designs
- there are some criteria for comparing designs

Therefore, there should have some relationships among the three kinds of designs. Fang et al. [13] found that many existing orthogonal arrays of strength two are uniform design under CD. Therefore, they proposed a conjecture that any orthogonal design is a uniform design under a certain discrepancy. Late, under the CD Ma, Fang and Lin [28] proved this conjecture is true for a full design q^s if $q = 2$ or q is odd, or $s = 1$ or 2. In general, the conjecture is not true. The study gives some relationship between orthogonality and uniformity.

3.2 Some criteria in experimental designs

Let us review some existing criteria and relationships among the criteria:

A. Minimum aberration and generalized minimum aberration: There are many useful criteria for comparing factorial designs, such as *resolution* (Box, Hunter and Hunter [2]) and minimum aberration (Fries and Hunter [18]). For given a regular factorial design D of s factors, its word-length pattern, denoted by $W(D) = (A_1(D), \dots, A_s(D))$, gives rich information on its statistical inference ability. A q^{s-k} regular FFD D is an $(s-k)$-dimensional linear subspace of q^s. The k-dimensional orthogonal subspace, denoted by $D^\perp$, of D is the *defining contrasts subgroup* of D. The elements of $D^\perp$ are called *words*. Let $A_i(D)$ be the number of distinct words of length i in the defining relation of D. Then the sequence $W(D) = \{A_1(D), \dots, A_s(D)\}$ is called the *word length pattern* of D. Ma and Fang [27] and Xu and Wu [42] independently extended the word length pattern to non-regular FFD. We still use $W(D) = \{A_1(D), \dots, A_s(D)\}$ for the generalized word length pattern. The *resolution* of D is the smallest i with positive $A_i(D)$ in $W(D)$. Let D_1 and D_2 be two designs. Let t be the smallest integer such that $A_t(D_1) \neq A_t(D_2)$

in their generalized word length patterns. Then D_1 is said to have less generalized aberration than D_2 if $A_t(D_1) < A_t(D_2)$. A design D has *minimum generalized aberration* (MGA) if no other q-level design has less generalized aberration than it. The MA/GMA is the most popularly used criterion in comparing FFDs.

B. $E(s^2)$ criterion: For a U-type design with two levels -1 and 1, let $\boldsymbol{X}$ be the design matrix where each row stands for the level-combination of a run and each column stands for a factor. Let s_{ij} be the (i,j)-element of $\boldsymbol{X}'\boldsymbol{X}$. The $E(s^2)$ criterion, proposed by Booth and Cox [1], is to minimize

$$E(s^2) = \sum_{1\le i<j\le s} s_{ij}^2 \Big/ \binom{s}{2}.$$

Obviously, $E(s^2) = 0$ for any orthogonal array, otherwise $E(s^2) > 0$. For any non-orthogonal design its lower bound was obtained by Nguyen [31]. Namely,

$$E(s^2) \ge \frac{n^2(s-n+1)}{(s-1)(n-1)}. \tag{3.1}$$

C. ave χ^2 criterion: For three-level supersaturated designs, Yamada and Lin [44] defined a measure for dependency between two factors $\boldsymbol{x}_i$ and $\boldsymbol{x}_j$ by

$$\chi^2(\boldsymbol{x}_i,\boldsymbol{x}_j) = \sum_{u,v=1}^{3} \frac{\left(n_{uv}^{(ij)} - n/9\right)^2}{n/9}, \tag{3.2}$$

where $\boldsymbol{x}_i$ and $\boldsymbol{x}_j$ are the ith and jth columns of $\boldsymbol{X}$, $n_{uv}^{(ij)}$ is the number of (u,v)-pairs in $(\boldsymbol{x}_i,\boldsymbol{x}_j)$. Then they defined a criterion for the whole design $\boldsymbol{X}$ by

$$\text{ave}\,\chi^2 = \sum_{1\le i<j\le s} \chi^2(\boldsymbol{x}_i,\boldsymbol{x}_j) \Big/ \binom{s}{2}.$$

ave $\chi^2 = 0$ for any orthogonal array, otherwise ave $\chi^2 > 0$. Yamada and Matsui [46] obtained a lower bound of ave χ^2 as follows:

$$\text{ave}\,\chi^2 \ge \frac{2n(2s-n+1)}{(n-1)(s-1)}. \tag{3.3}$$

More results can refer to Yamada et al. [45]

D. $E(f_{NOD})$ criterion: For a U-type design $U(n;q_1\times\cdots\times q_s)$ define

$$f_{NOD}^{ij} = \sum_{u=1}^{q_i}\sum_{v=1}^{q_j}\left(n_{uv}^{(ij)} - \frac{n}{q_iq_j}\right)^2, \tag{3.4}$$

where $n_{uv}^{(ij)}$ is the number of (u,v)-pairs of factor i and factor j and $n/(q_i q_j)$ stands for the average frequency of level-combinations in each pair of factors i and j. A criterion $E(f_{NOD})$ is defined as

$$E(f_{NOD}) = \sum_{1\le i<j\le s} f_{NOD}^{ij} \Big/ \binom{s}{2}.$$

Fang, Lin and Liu [12] found a lower bound for $E(f_{NOD})$

$$E(f_{NOD}) \ge \frac{n\Big(\sum_{j=1}^{m} n/q_j - m\Big)^2}{m(m-1)(n-1)} + C(n, q_1, \ldots, q_m), \tag{3.5}$$

where $C(n, q_1, \ldots, q_m) = \frac{nm}{m-1} - \frac{1}{m(m-1)}\left(\sum_{i=1}^{m} \frac{n^2}{q_i} + \sum_{i,j=1,\, j\ne i}^{m} \frac{n^2}{q_i q_j}\right)$ depends on the design only through $n, q_1, \ldots, q_m$.

E. Various discrepancies: We have introduced the star discrepancy defined in (2.5), the CD in (2.6) and the WD in (2.7). Hickernell and Liu [22] proposed the so-called the *discrete discrepancy* and Fang, Ge and Liu [6] found a lower bound for this criterion. The discrete discrepancy has played an important role for construction of uniform designs based on combinatorial designs. A comprehensive studies can refer to Fang et al. [6, 7, 8, 9].

3.3 Uniformity and word length pattern

The above criteria have been proposed by its own consideration and their lower bounds were obtained by different authors. Are there any relationships among these criteria? There are a lot of new development along this line. An important finding was by Fang and Mukerjee [15], where they obtained an analytic link between the CD and word-length pattern for any regular two-level factorials 2^{s-p} as follows:

$$[CD_2(D)]^2 = \left(\frac{13}{12}\right)^s - 2\left(\frac{35}{32}\right)^s + \left(\frac{8}{9}\right)^s \left\{1 + \sum_{i=1}^{s} \frac{A_i(D)}{9^i}\right\}. \tag{3.6}$$

Ma and Fang [26] extended the above result to the wrap-around L_2-discrepancy and three-level designs. They found relationships between WD and word-length pattern:

$$(WD_2(D))^2 = \begin{cases} \left(\frac{11}{8}\right)^s \sum_{r=1}^{s} \frac{A_r(D)}{11^r} + \left(\frac{11}{8}\right)^s - \left(\frac{4}{3}\right)^s, & \text{if } q=2, \\ \left(\frac{73}{54}\right)^s \left[1 + 2\sum_{j=1}^{s} \left(\frac{4}{73}\right)^j A_j(D)\right] - \left(\frac{4}{3}\right)^s, & \text{if } q=3. \end{cases} \tag{3.7}$$

The formulas (3.6) and (3.7) indicate that 1) the uniformity criterion is essentially consistent with the resolution and minimum aberration criteria;

2) the uniformity can be applied to both regular and non-regular factorial designs with any number of levels, but the resolution and minimum aberration can be applied only to regular designs in the past and designs with lower number of levels ($q = 2$ or 3 in most studies). Those results show the usefulness of the uniformity in factorial designs and some advantages of the uniformity. However, the discrepancy criteria has some weakness. Comparing the resolution criterion the discrepancy do not have such a clear criterion to classify designs into different levels. For overcoming this shortcoming Hickernell and Liu [22] proposed the so-called *projection discrepancy pattern* that has a similar function like the resolution. They indicated that the uniform design limit aliasing. Fang and Qin [16] proposed the *uniformity pattern* and related criteria for two-level designs, here the uniformity pattern likes the generalized word length pattern, but it is easy to computer.

3.4 Uniformity and isomorphism

Two U-type designs $U(n, q^s)$ are called *isomorphic* each other if one can be obtained from the other by relabelling the factors, reordering the runs, or switching the levels of one or more factors. For identifying two such designs a complete search must be done to compare $n!\,(q!)^s s!$ designs. Therefore, to identify the isomorphism of two $d(n, q, s)$ designs is known to be an NP hard problem when n and s increase. Ma, Fang and Lin [27] noted the fact that two isomorphic $U(n, q^s)$ designs should have the same uniformity and the same distribution of projection uniformity in all marginal subdimensions and proposed an efficient algorithm to detect non-isomorphism. Fang and Ge [5] extended the above idea for detecting inequivalence of Hadamard matrices and proposed a new algorithm. A Hadamard matrix, $\boldsymbol{H}$ say, of order n is an $n \times n$ matrix with elements 1 or -1, which satisfies $\boldsymbol{H}'\boldsymbol{H} = n\boldsymbol{I}$. Hadamard matrices have been played important roles in experimental designs and code theory. Two Hadamard matrices are called equivalent if one can be obtained from the other by some sequence of row and column permutations and negations. They applyed the new algorithm to Hadamard matrices of order 36 and discovered that there are at least 382 pairwise inequivalent Hadamard matrices of order 36. This was a new discovery.

3.5 Majorization framework

Recently, Zhang et al. [47] found a unified approach to describe the above criteria and their lower bounds by the use of the *majorization theory* (Marshall and Olkin [29]).

For two nonnegative vectors $\boldsymbol{x}, \boldsymbol{y} \in R_+^m$ with the same sum of its components. We write $\boldsymbol{x} \preceq \boldsymbol{y}$ if $\sum_{r=1}^k x_{[r]} \geq \sum_{r=1}^k y_{[r]}$, $k = 1, 2, \ldots, m-1$, where $x_{[1]} \leq x_{[2]} \leq \cdots \leq x_{[m]}$ are ordered numbers of x_i, and $y_{[r]}$ have the same meaning. A real-valued function ψ on R_+^m is called *Schur-convex* if

$\psi(\boldsymbol{x}) \leq \psi(\boldsymbol{y})$ for every pair $\boldsymbol{x}$, $\boldsymbol{y} \in R_+^m$ with $\boldsymbol{x} \preceq \boldsymbol{y}$. The summation/product of several Schur-convex functions is still a Schur-convex function, especially, an important class of Schur-convex functions are *separable convex* functions of the form $\Psi(\boldsymbol{x}) = \sum_{r=1}^m \psi(x_r)$ with $\psi''(x) \geq 0$.

For a U-type design $\boldsymbol{X}$ of n runs and s factors. Let d_{ij}^H be the *Hamming distance between runs i and k*. Let $\boldsymbol{d}(D)$ be m-dimensional *pairwise distance vector* (PDV) of d_{ij}^H for $1 \leq i < k \leq n$, where $m = n(n-1)/2$. For any design $U(n, q^s)$, the sum of its pairwise distance vector is uniquely determined by $d_{total} = \sum_{1 \leq i < k \leq n} d_{ij}^H = \frac{ns(n-q)}{2q}$. This fact gives possibility that we can apply the majorization theory to find a lower bound. Zhang et al. [47] found that the criteria, like $E(s^2)$, ave χ^2, A_2, A_3 in word length pattern, discrete discrepancy, wrap-around L_2-discrepancy ($q = 2$, 3) and centered L_2-discrepancy ($q = 2$) can be expressed as a separable Schur function of PDV.

Let $\bar{d} = d_{total}/m = \frac{s(n-q)}{q(n-1)}$, and let $\bar{\boldsymbol{d}}(D)$ be a $m \times 1$ vector of $\bar{d}$'s. From the majorization theory we have $\bar{\boldsymbol{d}}(D) \preceq \boldsymbol{d}(D)$. By this way we obtain a lower bound $m\psi(\bar{d})$, when $\bar{d}$ is an integer and the criterion has of the form $\Psi(\boldsymbol{x}) = \sum_{i=1}^m \psi(x_i)$. When $\bar{d}$ is not an integer, let $\bar{d}_i$ and $\bar{d}_f$ be the integral part and fractional part of $\bar{d}$, respectively. For any separable convex function $\sum_{i=1}^m \psi(x_i)$ it has a tight lower bound

$$m(1 - \bar{d}_f)\psi(\bar{d}_i) + m\bar{d}_f(\psi(\bar{d}_i) + 1). \tag{3.8}$$

Denote $\tilde{\boldsymbol{d}}(D)$ with the first $m(1 - \bar{d}_f)$ components of $\bar{d}_i$'s and following $m\bar{d}_f$ components of $1 + \bar{d}_i$'s. We have $\bar{\boldsymbol{d}}(D) \preceq \tilde{\boldsymbol{d}}(D) \preceq \boldsymbol{d}(D)$ by Lemma 5.2.1 of Dey and Mukerjee [3]. This approach gives a unified approach to find a lower bound for the criterion that can be expressed as a separable Schur function of PDV of a U-type design. When $\bar{d}$ is not an integer, the lower bound of (3.8) is new for all the criteria we have mentioned.

4 Applications of The Uniform Design

Since 1980 the uniform design has been widely used for various projects. For example, Ling, Fang and Xu (2001) gave a comprehensive review on applications of UD in chemistry and chemical engineering. There is a large potential applications of the uniform design for grammar of technology development, especially, for grammar of high tech development. Those who can first to develop a new high tech product and who will dominate the market, at least in the first few years. From the literature you can find the following about UD:

- There are more than 500 hundreds case studies published in more than one hundred journals.
- More than one hundred theoretic research papers have been published in various journals.

- Ford Motor Company has used UD for automobile development and "Design for Six Sigma".
- A nationwide society "The Uniform Design Association of China" was established in 1994 and has organized many public lectures, short courses, conferences and workshops.

The users appreciate the UD in the following aspects:

(a) flexibility in design and modeling;
(b) easy to understand and use;
(c) good for nonlinear models;
(d) can be applied on complicated system; and
(e) can be used for several occasions: physical experiments with unknown model, computer experiments, computer-based simulations and experiments with mixtures;
(f) computer aided software is available.

The Ford Motor Company has used the UD for developing new engines. Agus Sudjianto, Engineering Manager in FORD invited the first author of the paper to visit the FORD in 2002. His letter of invitation wrote: "In the past few years, we have tremendous in using Uniform Design for computer experiments. The technique has become a critical enabler for us to execute 'design for Six Sigma' to support new product development, in particular, automotive engine design. Today, computer experiments using uniform design have become standard practices at Ford Motor Company to support early stage of product design before hardware is available." It shows that there is a big potential applications of the uniform design in Six Sigma development. The monograph "Design and Modeling for Computer Experiment" by Fang, Li and Sudjianto presents many experiments that were implemented in the FORD.

Acknowledgement. The author would appreciate Prof. S. Yamada for his valuable comments and discussions. This research was partially supported by the Hong Kong RGC grant RGC/HKBU 200804 and Hong Kong Baptist University grant FRG/03-04/II-711.

References

1. Booth, K.H.V. and Cox, D.R. (1962) Some systematic supersaturated designs, *Technometric*, **4**, 489–495.
2. Box, G.E.P., Hunter, E.P. and Hunter, J.S. (1978) *Statistics for Experimenters*, Wiley, New York.
3. Dey, A. and Mukerjee, R. (1999) *Fractional Factorial Plans*, John Wiley, New York.
4. Fang, K.T. (1980) The uniform design: application of number-theoretic methods in experimental design. *Acta Math. Appl. Sinica*, **3**, 363–372.

5. Fang, K.T. and Ge, G.N. (2004) A sensitive algorithm for detecting the inequivalence of Hadamard matrices, *Math. Computation*, **73**, 843–851.
6. Fang, K.T., Ge, G.N. and Liu, M.Q. (2002) Uniform supersaturated design and its construction, *Science in China, Ser. A*, **45**, 1080–1088.
7. Fang, K.T., Ge, G.N, and Liu, M.Q. (2003) Construction of optimal supersaturated designs by the packing method, *Science in China (Series A)*, **47**, 128–143.
8. Fang, K.T., Ge, G.N., Liu, M. and Qin, H. (2004a) Construction of uniform designs via super-simple resolvable t-designs, *Utilitas Math.*, **66** 15–32.
9. Fang, K.T., Ge, G.N, Liu, M.Q. and Qin, H. (2004b) Combinatorial construction for optimal supersaturated designs, *Discrete Math.*, **279**, 191–202.
10. Fang, K.T., Li, R. and Sudjianto, A. (2005) *Design and Modeling for Computer Experiments*, Chapman & Hall/CRC Press, London.
11. Fang, K.T. and Lin, D.K.J. (2003) Uniform experimental design and its application in industry. *Handbook on Statistics 22: Statistics in Industry*, Eds. by R. Khattree and C.R. Rao, Elsevier, North-Holland, 131–170.
12. Fang, K.T., Lin, D.K.J. and Liu, M.Q. (2003) Optimal mixed-level supersaturated design, *Metrika*, **58**, 279–291.
13. Fang, K.T., Lin, D.K.J., Winker, P. and Zhang, Y. (2000) Uniform design: Theory and Applications, *Technometrics*, **42**, 237–248.
14. Fang, K.T. and Ma, C.X. (2000) The usefulness of uniformity in experimental design, in *New Trends in Probability and Statistics*, **Vol. 5**, T. Kollo, E.-M. Tiit and M. Srivastava, Eds., TEV and VSP, The Netherlands, 51–59.
15. Fang, K.T. and Mukerjee, R. (2000) A connection between uniformity and aberration in regular fractions of two-level factorials, *Biometrika*, **87**, 193–198.
16. Fang, K.T. and Qin, H. (2004) Uniformity pattern and related criteria for two-level factorials, *Science in China Ser. A. Mathematics*, **47**, 1–12.
17. Fang, K.T. and Winker, P. (1998) Uniformity and orthogonality, Technical Report MATH-175, Hong Kong Baptist University.
18. Fries, A. and Hunter, W.G. (1980) Minimum aberration 2^{k-p} designs, *Technometrics*, **22**, 601–608.
19. Hedayat, A.S., Sloane, N.J.A. and Stufken, J. (1999) *Orthogonal Arrays: Theory and Applications*, Springer, New York.
20. Hickernell, F.J. (1998) Lattice rules: how well do they measure up? in *Random and Quasi-Random Point Sets*, Eds. P. Hellekalek and G. Larcher, Springer-Verlag, 106–166.
21. Hickernell, F.J. (1999) Goodness-of-Fit Statistics, Discrepancies and Robust Designs, *Statist. Probab. Lett.*, **44**, 73–78.
22. Hickernell, F.J. and Liu, M.Q. (2002) Uniform designs limit aliasing, *Biometrika*, **389**, 893–904.
23. Liang, Y.Z., Fang, K.T. and Xu, Q.S. (2001) Uniform design and its applications in chemistry and chemical engineering, *Chemometrics and Intelligent Laboratory Systems*, **58**, 43–57.
24. Lin, D.K.L. (1993) A new class of supersaturated designs, *Technometrics*, **35**, 28–31.
25. Lin, D.K.J. (2000) Recent developments in supersaturated designs, in *Statistical Process Monitoring and Optimization*, Eds. by S.H. Park and G.G. Vining, Marcel Dekker, Chapter 18.
26. Ma, C.X. and Fang, K.T. (2001) A note on generalized aberration in factorial designs, *Metrika*, **53**, 85–93.

27. Ma, C.X., Fang, K.T. and Lin, D.K.J. (2001) On isomorphism of fractional factorial designs, *J. Complexity*, **17**, 86–97.
28. Ma, C.X., Fang, K.T. and Lin, D.K.J. (2003) A note on uniformity and orthogonality, *J. Statist. Plan. Inf.*, **113**, 323–334.
29. Marshall, A.W. and Olkin, I. (1979) *Inequalities: Theory of Majorization and Its Applications.* Academic Press, New York.
30. Niederreiter, H. (1992) *Random Number Generation and Quasi-Monte Carlo Methods*, SIAM CBMS-NSF Regional Conference Series in Applied Mathematics, Philadelphia.
31. Nguyen, N.-K. (1996) An algorithmic approach to constructing supersaturated designs, *Technometrics*, **38**, 69–73.
32. Sacks, J., Welch, W.J., Mitchell, T.J. and Wynn, H.P. (1989) Design and analysis of computer experiments. *Statistical Science.* **4**, 409–435.
33. Santner, T.J., Williams, B. and Notz, W.I. (2001) *The Design and Analysis of Computer Experiments*, Springer, New York.
34. Saltelli, A. and Chan, K. and Scott, E.M. (2000) *Sensitivity Analysis*, Wiley, New York.
35. Stein, M.L. (1999) *Interpolation of Spatial Data, Some Theory for Kriging*, Springer, New York.
36. Wang, Y. and Fang, K.T. (1981) A note on uniform distribution and experimental design, *KeXue TongBao*, **26**, 485–489.
37. Wiens, D.P. (1991) Designs for approximately linear regression: two optimality properties of uniform designs, *Statist. & Prob. Letters.*, **12**, 217–221.
38. Winker, P. and Fang, K.T. (1997) Application of Threshold accepting to the evaluation of the discrepancy of a set of points, *SIAM Numer. Analysis*, **34**, 2038–2042.
39. Winker, P. and Fang, K.T. (1998) Optimal U-type design, in *Monte Carlo and Quasi-Monte Carlo Methods 1996*, eds. by H. Niederreiter, P. Zinterhof and P. Hellekalek, Springer, 436–448.
40. Wu, C.F.J. and Hamada, M. (2000) *Experiments: Planning, Analysis, and Parameter Design Optimization*, Wiley, New York.
41. Xie, M.Y. and Fang, K.T. (2000) Admissibility and minimaxity of the uniform design in nonparametric regression model, *J. Statist. Plan. Inference*, **83** 101–111.
42. Xu, H. and Wu, C.F.J. (2001) Generalized minimum aberration for asymmetrical fractional factorial designs. *Ann. Statist.*, **29**: 1066–77.
43. Yamada, S. and Lin, D.K.J. (1997) Supersaturated design including an orthogonal base, *Canad. J. Statist.*, **25**, 203–213.
44. Yamada, S. and Lin, D.K.J. (1999) Three-level supersaturated designs. *Statist. Probab. Lett.* **45**, 31–39.
45. Yamada, S., Ikebe, Y.T., Hashiguchi, H. and Niki, N. (1999) Construction of three-level supersaturated design, *J. Statist. Plann. Inference*, **81**, 183–193.
46. Yamada, S. and Matsui, T. (2002) Optimality of mixed-level supersaturated designs, *J. Statist. Plann. Infer.* **104**, 459–468.
47. Zhang, A.J., Fang, K.T., Li, R. and Sudjianto, A. (2005) Majorization framework balanced lattice designs, *The Annals of Statistics*, **33**, 2837–2853.

Adapting Response Surface Methodology for Computer and Simulation Experiments

G. Geoffrey Vining

Virginia Tech, Blacksburg, Virginia

1 Abstract

Response surface methodology (RSM) is a powerful experimental strategy for optimizing products and processes. RSM originated in the chemical process industry (CPI), but it has found widespread application beyond the CPI, including finance. Traditional RSM is a sequential learning process that assumes a relatively few number of important factors. It also assumes that low order Taylor series approximations are appropriate over localized experimental regions.

Engineers and scientists are using computer and simulation models as the basis for product and process design. These models are extremely complex and typically highly nonlinear. Analytic solutions do not exist. As a result, computer experiments are becoming an important basis for optimizing such systems.

Computer experiments present several challenges to RSM. First, the underlying models are deterministic, which means there is no random error. In some cases, random errors are introduced, especially in simulation experiments. Nonetheless, the models themselves are deterministic. Second, these experiments often involve a relatively large number of potential factors. Traditional RSM designs prove much larger than can actually be conducted.

This paper reviews the application of RSM to computer and simulation experiments. It examines when traditional RSM should work well. It then explores adaptations of RSM for computer and simulation experiments.

2 Introduction

For many years, practitioners have applied response surface methodology (RSM) to optimize physical products or processes. RSM provides a powerful set of tools that under the proper circumstances often has yielded significant product and process improvements.

Recently, engineers have begun using extremely complex mathematical models that closely mimic products and processes that are too expensive to do extensive physical experimentation, for example aircraft wings. The computer codes required for these complex mathematical models are themselves expensive to run although clearly nowhere near as expensive as the physical product or process. Engineers are turning to experimental design strategies to interact with these complex mathematical models in a more efficient manner.

In other cases, engineers are using complex simulation models to study phenomena of interest, for example an inventory system. Once again, the goal is to find optimal conditions.

RSM with its emphasis on product and process optimization seems a logical candidate strategy for both these computer experiments and these simulation studies. However, there are subtleties to both computer and simulation experiments that complicate the direct application of RSM.

This paper gives a general overview of the application of RSM to computer and simulation experiments. This paper proceeds first with a basic introduction to RSM. It next presents a brief overview of computer experiments. The next section contrasts RSM and computer experiments. This paper concludes with a few suggestions for adapting RSM to computer experiments.

3 Introduction to RSM

RSM originated in the chemical processes industry and has been successfully applied to many complex products and processes. RSM seeks "optimal" product or processing conditions. RSM's basic strategy is to use sequential experimentation. The classic texts describing RSM are Box and Draper [1], Khuri and Cornell [2], and Myers and Montgomery [3].

The seminal paper for RSM was Box and Wilson [4]. This paper outlines an sequential approach for optimizing a process based on low-order Taylor series approximation. On many occasions, Box has said, "All models are wrong; some models are useful." In the early phases of RSM lack-of-fit tests are important for making decisions about model transition. It is important to note that RSM assumes only a few "active" factors. The product or process may be extremely complex with many possible factors. However, RSM assumes enough previous knowledge, generally based on engineering theory, to concentrate on a relatively few potential factors.

Figure 1 gives an overview of the sequential nature of RSM. The region of operability represents all feasible settings for the product or process factors. Typically, the region of operability is too large to assume low order Taylor series approximations for the characteristic of interest, hence, the need for sequential experimentation.

The paradigm for RSM is the transition from bench-top to pilot plant operation or from pilot plant to full scale operation. Within this paradigm, the practitioner has a recommended set of operating conditions that should

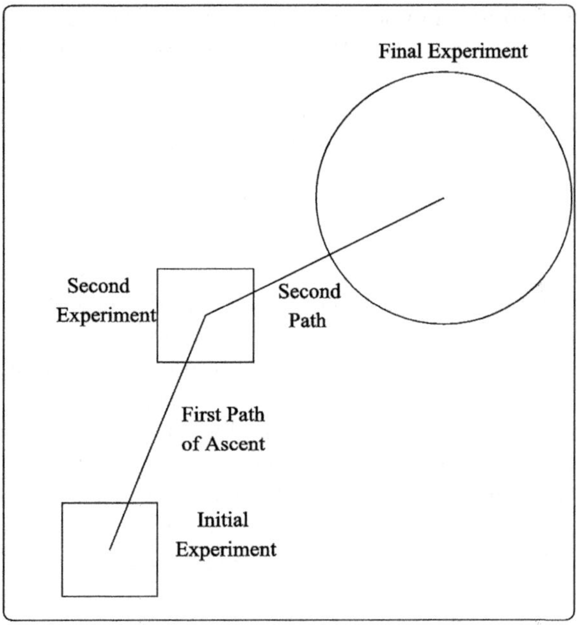

Fig. 1. Outline of Sequential Experimentation in RSM

be relatively well specified for many of the potential factors but almost certainly some of the factors are poorly specified, with some perhaps grossly misspecified.

RSM begins with a small experiment in the neighborhood of the recommended set of operating conditions. The experimenter must determine the set of factors and their levels for this experiment. At this point the experimenter assumes a strict first-order model for the response. The two primary purposes of this experiment are to screen out the inactive factors and to find the direction for better results. It is crucial to use as small an experimental design as possible at this stage in order to save resources for the latter stages of experimentation.

There are several possible follow-up experiments. The choice of follow-up experiment depends upon what is learned from the first experiment and upon subject area expertise.

Most commonly, the experimenter conducts a path of steepest ascent (or descent, depending on the objective). Based on the estimated model, the experimenter determines a path that, according to the model will produce better results. The idea then is to conduct a series a runs along this path until he/she no longer sees improvement.

In some cases, the experimenter really is in the neighborhood of the optimum conditions. The lack-of-fit test is important for making such a determination. In this situation, the appropriate follow-up experiment is to augment the original design to accommodate a second-order design. It is rare to perform such an augmentation after the very first experiment although it can occur.

Other options for follow-up experiments include dropping some factors while adding new factors as well as changing the levels of certain factors and then running a second experiment in the same neighborhood as the first. Experimenters rarely exercise this option after the very first experiment, but it does occur.

Typically, the experimenter pursues a path of steepest ascent away from the original experimental neighborhood or region. Once the path leads to no more improvement, the experimenter conducts a new experiment in the neighborhood of the best conditions encountered to this point. The ability to perform lack-of-fit tests is very important to these follow-up experiments. These tests can point out the presence of significant curvature, which indicates the neighborhood of at least a local optimum.

Ultimately, the series of experiments leads the experimenter into the neighborhood of at least a local optimum. At this point, he/she conducts a second-order optimization experiment. Typically, the experimenter has learned enough about the system to focus on just a few active factors, which is important because second-order designs are rather large. The second-order model affords a great deal of insight about what affects the response near the optimum. This insight often proves invaluable for later operation of the particular product or process. Ideally, the experimenter conducts a confirmatory experiment in the neighborhood of the recommended optimum.

In summary, RSM is truly a sequential learning strategy. Properly applied, it does not use a single, one-shot experiment. Rather, each experimental phase is based on the previous experimentation. The true goal of RSM is to find better operating conditions and not necessarily the absolute best. As a result, converging to a local optimum is often quite acceptable. To be successful, RSM needs a relatively good starting point (initial recommended operating conditions) and relatively good information about the likely "active" factors. **It is important to note that an active factor is one that is very important to the response and grossly misspecified in the initial recommendation.** RSM assumes a relatively small number of active factors and a relatively smooth surface. In such an environment, a series of experiments based on low-order Taylor series approximations proves to be a powerful strategy for improvement.

Properly applied, RSM provides significant insight about what controls the response near the optimum operating conditions. To be successful, RSM requires

- thoughtful interaction with the process,
- good initial information, and
- intelligent choice of follow-up experimentation.

It is crucial in RSM to keep the project manageable.

4 Computer Experiments

Historically, the key to making scientific and engineering advances has been the scientific method, which involves a constant interaction between a concrete problem and an abstract representation of that problem. The scientific method uses mathematical theory to model physical phenomena. For many years, the scientific method sought analytical solutions because these mathematical models were relatively simple. For many modern interesting scientific and engineering problems, there are no analytical solutions because the mathematical models require extensive computation.

In computer and simulation experiments, extremely complex mathematical models become the "underlying reality" for our problems. These experiments require an appropriate strategy for interacting with the mathematical models. Unfortunately, too often people forget the adage that "All models are wrong; some models are useful." These complex mathematical models are themselves imperfect surrogates for the real physical phenomena. In some cases, these models produce surfaces far "bumpier" than reality due to constraints and certain artifacts in the modeling procedure. A response surface can be rather bumpy in the sense of many possible local optima even if all of the functions are continuously differentiable. Bumpy response surfaces present significant challenges for optimization. A mathematical model that is bumpier that the real system compounds this problem.

This paper distinguished between strict computer experiments and simulation experiments. For strict computer experiments, the underlying models are deterministic. The possible "error" is due to round-off. There is no real "noise." Running the same levels for the factors will produce the same result each time since the models are deterministic. Traditional RSM assumes the presence of true background noise. Hence, one should expect some issues when adapting RSM to these kinds of computer experiments.

Simulation experiments start with deterministic models and then add errors to the inputs to produce noise in the response. Traditional RSM works better for these simulation experiments, but there are still issues.

Santer, Williams, and Notz [5] provide a nice overview of computer experiments from a statistical perspective. Both computer and simulation experiments assume that the experimental region is in fact the entire region of operability. In such a situation, low-order Taylor series approximations typically are no longer appropriate. The response surfaces over the entire region of operability tend not to be smooth. By smooth, we mean having few local optima. Models are often difficult to identify. In many cases, the analysts makes no attempt to fit a specific model. Instead, he/she uses some form of nonparametric modeling as the basis for prediction. In such cases, some analysts resort to "pick the winner" optimization procedures. Typically, computer experiments offer limited information, at best, about what affects the response near the optimum operating conditions.

Computer and simulation experiments often involve a large number of possible factors, which leads to an extremely high dimensional design space. In addition, the experimenter often has poor initial information about the starting point and about the active factors. Typically, computer and simulation experiments are one-shot and not sequential. The absence of true noise makes statistical tests difficult either to perform or to interpret.

Uniform designs are very popular one-shot approaches for computer and simulation experiments. Under appropriate conditions, uniform designs are "optimal" for non-parametric regression. One can employ a deterministic search strategy such as the genetic algorithm or kreiging in conjunction with a uniform design to optimize the product or process.

5 RSM and Computer Experiments

Table 1 contrasts traditional RSM and computer experiments. It is pretty clear that the usual approaches to computer experiments are quite different from the traditional strategy of RSM.

Table 1. Contrast of RSM and Computer Experiments

RSM	Computer Experiment
Series of Experiments	One-Shot Experiment
Few Active Factors	Many Factors
Small Design Space	Large Design Space
Low Dimension	High Dimension
Subset of Region of Operability	Entire Region of Operability
Assumes Relatively Smooth Surface	Surface Often Relatively Bumpy
Well Defined Models	Often No Real Model
True Background Noise	No or Artificial Background Noise
Describes Behavior Near Optimum	Often Cannot Describe Behavior Near Optimum

It is interesting to note that some people have used second-order, response surface designs as computer experiments. However, these applications have tended to be one-shot experiments over the entire region of operability. In many cases, these applications had little or poor initial information about the starting point and about the active factors. In these applications, RSM has tended to converge to unsatisfactory local optima. The analysts have been frustrated by their limited ability to perform statistical tests and their inability to perform lack-of-fit tests. In light of Table 1, it should surprise very few people that traditional RSM can fail to work satisfactorily in many computer experiment applications.

It is clear that many past applications of RSM to computer experiments were not appropriate. Traditional RSM can work well if the analyst has good information about the starting conditions. In such a case, the starting point should be in some reasonable neighborhood of an acceptable optimum. There must be good information about the potential active factors. Further, there should be only a relatively few potential active factors. RSM in a computer experiment must use steepest ascent as the basis for transition to the second-order model instead of formal lack-of-fit tests. The key to the transition is when steepest ascent no longer yields any improvements, which is an indication that the experiment is in the neighborhood of at least a local optimum. In all honesty, traditional RSM tends to work better for simulation experiments, which do have some noise, than computer experiments.

6 Modifying RSM

An interesting question is how to modify RSM for computer experiments. The key underlying element for modifying RSM is to recognize that at its heart RSM is a sequential learning strategy.

One proper way to adapt RSM to computer experiments is to use RSM ideas to modify deterministic search strategies. Such an idea is an extension of simplex evolutionary operation (EVOP). Each stage consists of one or more runs. One can use non-parametric and semi-parametric modeling techniques to avoid "pick the winner" optimization strategies. Traditional RSM could provide a reasonable basis for conducting a confirmatory experiment. The RSM confirmatory experiment would provide insight into the nature of the product or process in the neighborhood of the optimum.

Other approaches for adapting RSM could borrow certain Bayesian ideas that formally incorporate prior knowledge. Such approaches would formalize the sequential learning process. The analyst would make decisions based on posterior distributions.

In summary, traditional RSM has been an important tool for optimizing many physical products and processes. It is logical to assume that RSM would have potential as an important tool for computer experiments. It is important to note that a naive application of traditional RSM to computer experiments is fraught with peril. However, with some creative thought, we should expect adaptations of RSM to work well.

Clearly, computer and simulation experiments form an important part of the grammar of technology development. The modern scientific method depends heavily upon extremely computational approaches for modeling complex physical behavior. Computer and simulation experiments are essential for interacting efficiently with these computational models. As a result, computer and simulation experiments address core issues for future technology development. It is difficult to imagine a grammar of technology development that does not feature prominently computer and simulation experiments.

References

1. Box, G.E.P. and Draper, N.R. (1987) *Empirical Model-Building and Response Surfaces.* John Wiley & Sons, New York, NY.
2. Khuri, A.I. and Cornell, J.A. (1996) *Response Surfaces.* Marcel Dekker, New York, NY.
3. Myers, R.H. and Montgomery, D.C. (2002) *Response Surface Methodology: Process and Product Optimization Using Designs Experiments*, 2nd ed. John Wiley & Sons, New York, NY.
4. Box, G.E.P. and Wilson, K.B. (1951) On the experimental attainment of optimum conditions, Journal of the Royal Statistical Society, Ser. B, 13, pp. 1–45.
5. Santer, T.J., Williams, B.J. and Notz, W.I. (2003) *The Design and Analysis of Computer Experiments*, Springer, New York, NY.

SQC and Digital Engineering

Technological Trends in Design Parameter Optimization and Current Issues

Mutsumi Yoshino[1] and Ken Nishina[2]

[1] DENSO Corporation
yoshino@prd.denso.co.jp
[2] Nagoya Institute of Technology
nishina@nitech.ac.jp

Summary. Simulation experiments in Computer Aided Engineering (CAE) has been developing tecnologies. The application of Design of Experiments (DOE) to simulation experiments, which derives optimum parameters efficiently, is becoming as popular as actual experiments. However, some statistics point out that there are many misuses of the application of DOE in CAE. One of the main reasons is the lack of systematic procedure for it. Specifically, while technical development in the optimization problems is crucially required, the practice of optimization generally depends on experimental or institutional skills of the CAE engineers. Regrettably, the latent optimization problem has not been fully discussed. This paper focuses on it and aims to obtain a structure by reviewing the practical procedures. The final report of this paper is summarized in a table that suggests a direction of effective application of DOE in CAE.

Key words: Design parameter optimization, Multiobjective optimization, Robust optimization, Design of experiments, Response surface

1 Introduction

From the early developmental stage of SQC, various types of design parameter optimization have been undertaken through the SQC method. The most common procedure was to select the optimal level according to the results of an experiment that allocates a set of design to the multi-level design of experiments. However, a combination of response surface modeling and the operations research method has also been employed to estimate the optimum value.

* Contributed paper: received date: 17-Jul-05, accepted: 27-Feb-06
** This is a revised version of our earlier paper in the Journal of the JSQC, Vol.34, No. 3, pp. 5-12 (in Japanese).

The recent advancement in optimization software utilizing response surface methodology has enabled simultaneous optimization of multiple-linebreak responses. This evolved into robust optimization featuring simultaneous optimization of responses of mean values and errors vis-à-vis the characteristic value. In this paper, some of the recent optimization methods are reviewed. In addition, the peculiarities of characteristic values derived from simulations are highlighted in order to present current issues inherent in the modern optimization technologies.

2 Trends in Optimization Technology

Today, the design environment is increasingly computerized through the use of CAD (Computer-Aided Design). Moreover, CAD is becoming progressively integrated with CAE (Computer-Aided Engineering) and CAM (Computer-Aided Machining). The link between CAD and CAE enables automatic creation of finite element meshes and instantaneous analysis in the design stage. In other words, this link has contributed to on-the-spot and on-screen detection of defects, including resonance and a lack of strength. The performance of finite element analysis software (hereinafter referred to as "solver") has been improving and solves a variety of problems.

Recently, the growing popularity of digital engineering in the design environment resulted in the emergence of a new type of software called an optimization support engine. This software is capable of allocating design parameters to the design of experiments for analysis as well as controlling a solver directly. As a result, trial and error efforts by design engineers were replaced with the computers that can determine the optimum value automatically by changing the design. Computerization has enabled the optimum value to be calculated through various computerized SQC methods, thereby taking over the role of skillful design engineers who have extensive knowledge of the design of experiments. The latest trends in optimization are charactarized by the fact that for the computation and analysis of characteristic values, more CAE-based simulations are employed than actual experiments using conventional SQC methods.

According to Oda et al. [1], until approximately 1990, optimization support engines were only applicable to the operation research methods or the optimized standard methods. In such an environment, an operation research method routine operated the solver directly in an attempt to converge on the optimum value. This means that the software was only significant in eliminating trial and error efforts from design engineers. Around 1995, demand for an efficient way of optimum point search began to increase. Kawamo et al. [2] introduced an approach wherein the design parameters are allocated to the design of experiments, and then the responses of characteristics approach a response surface function with a small number of runs. Consequently, the optimum value is determined efficiently. At the 2000 meeting of the Optimization

Symposium (OPTIS), held biennially since 1994 by the Japan Society of Mechanical Engineers, approximately 25% of the presentations focused on optimization cases based on response surface approximation: this confirms the fact that this approach now prevails in the field of optimization.

In 1998, the Japan Society of Mechanical Engineers presented an award to Kashiwamura and Shiratori et al. [3] for their robust optimization research. This appears to indicate the growing popularity of robust optimization in recent years.

In 1995, Dr. Tong, an accomplished research fellow at General Electric Company, had created i-SIGHT, which is the epitome of optimization support engines, and later he established his own company to market his product. i-SIGHT operates within a mainframe computer and is rapidly getting prevalent despite its relatively high annual licensing fee. A reasonably-priced optimization software that operates on a PC is also widely used, though it lacks the solver control function.

Since the 2000 symposium, the Quality Engineering Society has held regular conferences titled "Simulation-based Robust Design." The Taguchi Methods have changed in response to the digital engineering age; the ideal functions had been expressed by a linear model in the Methods. However, as explained in more detail later, in simulations, variations of error value only pertain to nonlinear characteristics. To solve this problem, the Methods applied a nonlinear model called the standard S/N ratio.

3 Types of Optimization Problems

Figure 1 shows optimization problems categorized into single-objective, multi-objectives, and multidiscriplinary optimization problems. Single-objective optimization problem is to find the optimum parametric settings by measuring the responses of the system and adjusting them to the target value. Multi-objective optimization problem refers to when there is only one system to be optimized but the system has multiple responses. If responses are dealt with as mean values and errors, this problem involves robust optimization. Multidisciplinary optimization problem is defined as the optimization of a large-scale system, which makes its problem complicated. For, as illustrated, a response from a certain system acts as a design parameter of another system. Yamakawa [4] reports that recently this has been the focus of much attention in the American Society of Mechanical Engineers (ASME).

4 Optimization Methods

4.1 Case Study Outline

In this case study, the design parameters of an RC (radio control) car are optimized. Table 1 list 19 parameters and their respective value ranges. The

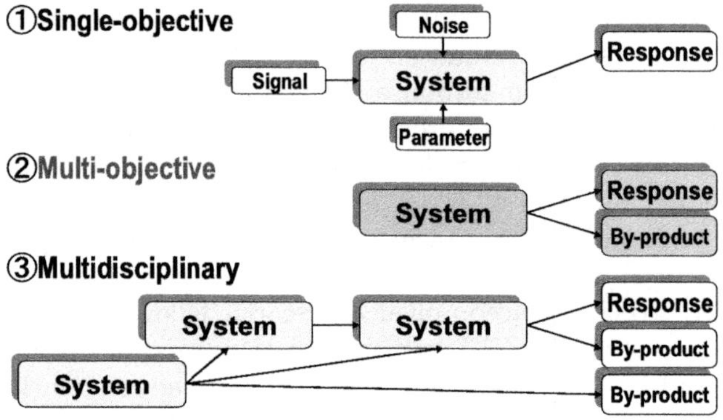

Fig. 1. Category of Optimization

characteristic value in this experiment is the lap time, or the time required to complete one circuit of the test course. The optimization problem is how to determine each parameter value so as to minimize the lap time. Simulators are available for download on the Internet [5]. In order to create a response surface function, the coefficients and constants terms for the linear effect (main effect), nonlinear effect, and two-factor interactions must be determined with respect to each parameter. Accordingly, when the number of parameters is defined as p, the number of equations, that is, the number of independent analyses required (N), is expressed in the equation (1) below:

$$N = p + p + {}_pC_2 + 1 = (p+1)(p+2)/2 \quad (1)$$

If $p = 19$, a total of 210 analysis runs are required. To avoid such an ineffective analysis, the parameters with the most significant influence on the responses should be selected prior to the analysis: this step is called screening. When the number of parameters is limited to $p = 5$, according to the equation (1) only 21 runs are necessary.

Following screening, a response surface is also widely used, though it determined. The parameters in the experiment should be allocated appropriately so that each coefficient is divided distinctly in approximately 21 runs. The lap time determined according to the allocation of the designed experiment is then entered into the optimization software to calculate a response surface. The shortest lap time occurs when the response surface reaches the deepest point in the vertical direction. Therefore, the parametric value at this point is the optimum value for the design.

The optimum value can be calculated in the procedure mentioned above. However, this procedure is inadequate as an actual design step. Response surfaces near the optimum value include a region with a steep gradient, as shown in Figure 2, and with a relatively flat one, as shown in Figure 3. In a steep surface region, parameter dispersion results in a drastic deterioration of characteristic value stability. On the other hand, the characteristic value tends

Table 1. Parameter List of RC-car

Chassis parameter	Abbr.	MIN	MAX	MEAN
Weight (kg)	SZ	1.2	1.8	1.5
Tire grip (G)	TG	1.28	1.92	1.6
Weight ratio of traction wheels	KH	0.8	1.2	1
Diameter of driving wheel (mm)	KC	56	84	70
Gear ratio	GR	2.0	6.0	4.0
Gear efficiency	GK	0.68	1.02	0.85
Rolling resistance coefficient	KT	0.0528	0.0792	0.066
Apparent weight of rotation (kg)	KS	0.18	0.27	0.225
Loading ratio of braking wheel	SK	SK = KK		
Decrease of load on braking wheel	SU	0.56	0.84	0.7
Loading ratio on front wheel	ZK	ZK = 1 − KK		
Loading ratio on rear wheel	KK	0.4	0.6	0.5
Aerodynamic parameter	**Abbr.**	**MIN**	**MAX**	**MEAN**
Drag coefficient (CD)	CD	0.294	0.788	0.541
Front projection area (m^2)	ZT	0.0191	0.0224	0.02075
Lift coefficient on braking wheel (Cl)	SD	SD = KD		
Lift coefficient on front wheel (Clf)	ZD	−0.032	0.186	0.077
Lift coefficient on rear wheel (Clr)	KD	0.082	0.84	0.461
Motor Power	**Abbr.**	**MIN**	**MAX**	**MEAN**
Maximun speed (rpm)	MR	16500	23700	20100
Maximum torque ($kg \cdot cm$)	MT	1.73	1.98	1.855

to remain stable in a flat surface region, even with a slight degree of parametric dispersion. A flatter slope is clearly more acceptable; thus, it is important to stabilize characteristic value responses against parametric variation. This is the definition of a robustness strategy.

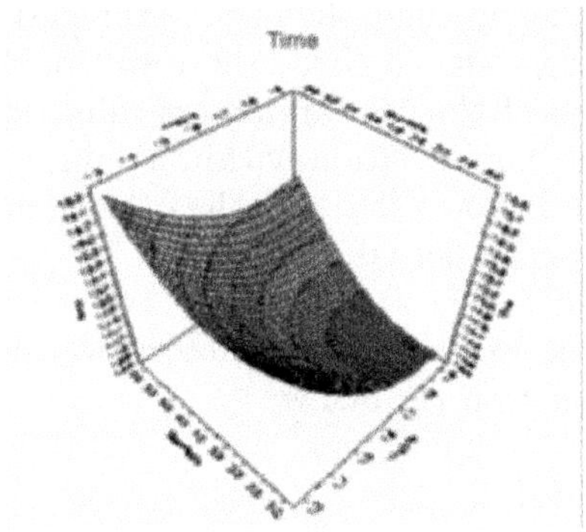

Fig. 2. Sharp Response

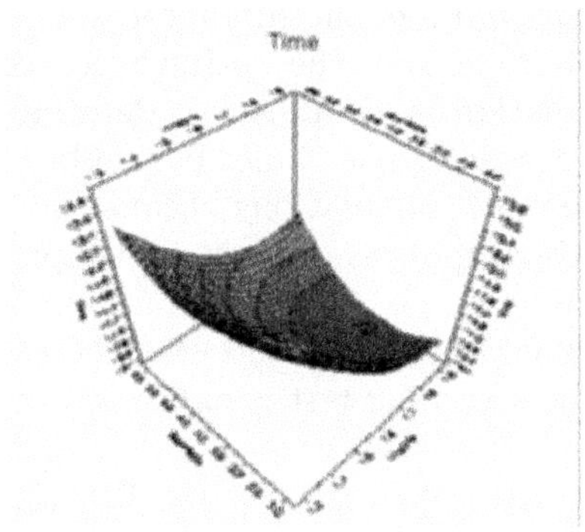

Fig. 3. Flat Response

In the following sections, a variety of SQC methods for optimization will be reviewed step by step. The optimization software used in this study is a PC-based program entitled MODDE (INFOCOM, Corp).

4.2 Screening of Design Parameters

Screening aims to determine whether parameters have an effect on the characteristic value in question. Because it is sufficient to resolve only the linear effect in screening, the Resolution III two-level design of experiments is employed. To prevent design engineers from failing to detect significant parameters due to a convex effect, center-point level runs for all parameters are often added to this design as a precautionary measure. In simulations, a run at the center-point level will suffice at one repetition since a simulation yields no random errors.

Because some of the parameters used in this case study already have functional relationships with each other, they are excluded from the observation: the remaining 16 parameters are evaluated in terms of their effects. When these parameters are defined and the optimization software operates screening, the software suggests several possible designs of experiments, in which the most recommended option is underlined, as shown in the following example:

Design	Recom.	Runs	Model
Frac Fac Res V		256	Interaction
D-Optimal		144	Interaction
Frac Fac Res IV	First	32	Linear
Plackett Burman	Second	20	Linear
D-Optimal		32	Linear

The Resolution IV design of experiments requires 33 runs, including a center-point level run. It is possible to downscale the experiment, but the lowest parameter is a confounding default. In accord with this rule, we therefore use the recommended design of experiments.

The optimization software creates an allocation table, which is then cut and pasted into the simulator to calculate the lap time; then the optimization software goes to the optimization-computing routine. Next, the result of the calculation is examined to determine which of the 16 parameters influences the characteristic value. It should be noted that variation within a subgroup is absent in simulations because no repeat error occurred. On the other hand, ANOVA analysis of variance is applicable for an actual experiment.

Where x_i is defined as a normalized value for each parameter, ξ as the lack of fit for the model and ε as the experimental error, the characteristic value in the screening test is expressed by the equation (2) below:

$$y = \sum \beta_i x_i + \xi + \varepsilon \tag{2}$$

β_i is the effect of each parameter. In simulations, $V(\varepsilon) = 0$. Therefore, the variance ratio between ε and β_i is incomputable, thus in principle eliminating the possibility of performing ANOVA. Since screening does not take interactions into consideration, the lack of fit due to these interactions is included in the term ξ. Thus, it is possible to compare the size of β_i descriptively in relation to ξ. However, it should be noted that the level of ξ varies depending on the scale of the design of experiments.

On the other hand, it is impossible to pool the effects of insignificant parameters in the error term ε in order to perform ANOVA. This is because, in simulations, every single parametric value is incorporated in the calculation request. No matter how small its effect may be, each parameter value is by no means undetectable nor free of deviation. As a result, even if ε is determined by pooling these insignificant effects, the outcome does not conform to normal distribution. For this reason, the F value derived from such pooling should not be used as the basis of evaluation of statistical significance. Unfortunately, there is no alternative means of evaluating statistical significance at present.

Here, the top five parameters are selected by comparing the absolute value of β_i in a Pareto chart. As Figure 4 shows, TG, GR, MR, KC and KT are regarded as having a major effect on the output. To ensure the accuracy of this selection, each distribution chart is examined to avoid overlooking important parameters due to a convex effect. In this case study, there are no convex effects other than the parameters selected above.

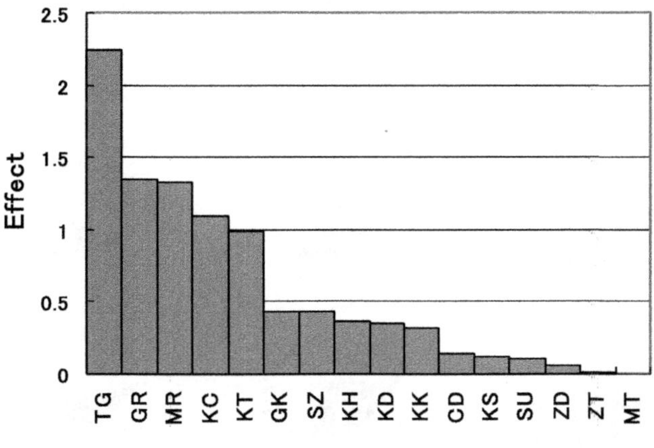

Fig. 4. Pareto Chart of Effect

4.3 Response Surface Modeling

These five major parameter levels are then varied within a certain range to create a response surface. The other parameter levels are fixed at the center level, since their influence on the characteristic value response is negligible. The Resolution V design of experiments is used for Response Surface Modeling (hereinafter referred to as "RSM"). A three-level design of experiments is required to estimate the nonlinear effects.

When the optimization software operates RSM, the software suggests several possible designs of experiments, in which the most recommended option is underlined, as shown in the following example.

Design	Recom.	Runs	Model
Full Fac (3 levels)		243	Quadratic
Box Benhken		40	Quadratic
CC (rotatable)		26	Quadratic
CC (face centered)	First	26	Quadratic
D-Optimal	Second	26	Quadratic

The Centered Composite Design requires 27 runs, including a center-point level run. The optimization software creates an allocation table in the same procedure as screening, which is then cut and pasted into the simulator to calculate the lap time. Then the software immediately creates a response surface model.

Variable screening, based upon regression analysis, is conducted at this stage to eliminate unnecessary interactive and nonlinear terms. The details of this process are omitted in this paper, due to space limitations. The response surfaces created are shown in Figure 5. The parametric setting that minimizes the distance between the response surface and the bottom is the optimum, unless robust optimization is performed.

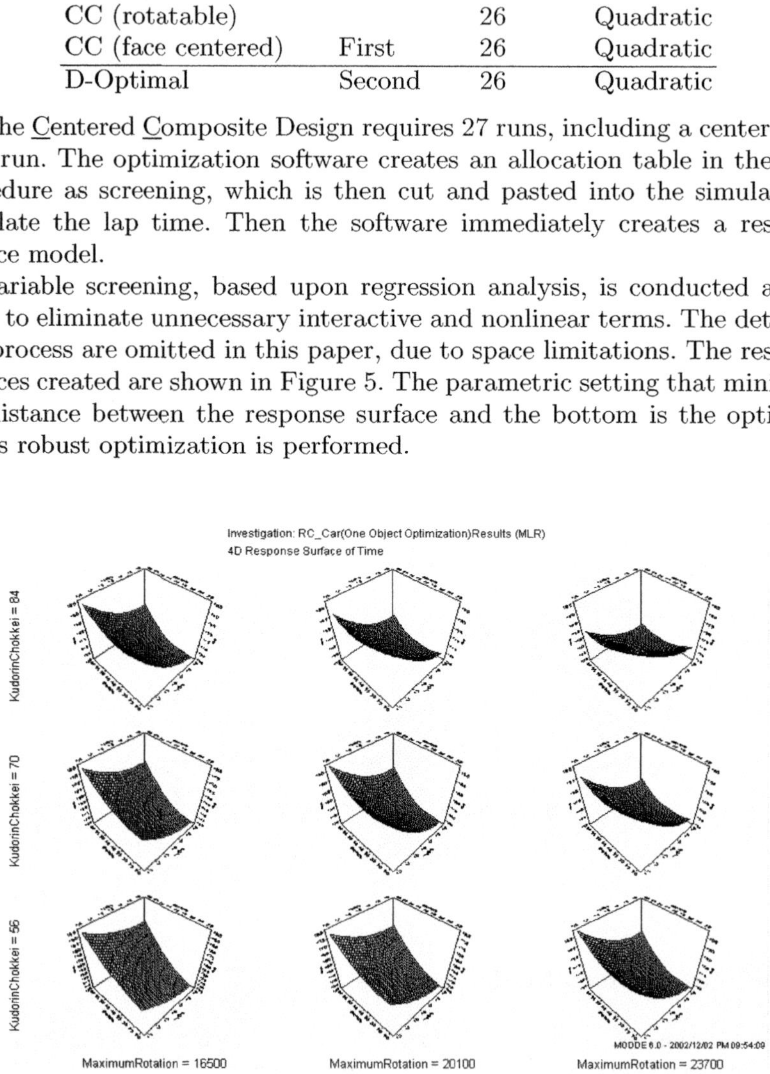

Fig. 5. Response Surface of Lap-time (axis Right; TG–Left; GR, array Horizontal; MR–Vertical; KC)

4.4 Robust Optimization

In robust optimization, a statistic of the characteristic value deviation (i.e. the range and standard deviation) is calculated and defined as an error. It is then optimized along with the characteristic value response that has been computed in the preceding process. This method might be criticized in terms that it presumes a model for an error that is intrinsically a random factor. However, since this method aims to measure the responses of the deviation by assigning a range of parametric values in the response surface approximate equations, as illustrated in Figure 6, it is acceptable to express the error as a function of the parameters.

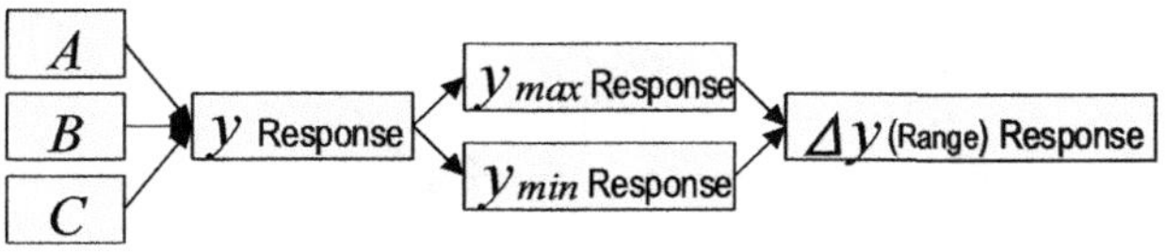

Fig. 6. Measurement of Error Response

Since responses of mean values and errors vis-à-vis characteristic values are derived from a common function and are thus dependent on each other, multi-objective simultaneous optimization is performed based upon the "Dual Response Modeling Approach" suggested by Myers et al. [6]. The crucial point in this step is to approximate the errors with the design parameter function. Akagi [7] maintains that the error should be first estimated by partial differentiation of the response surface function and a subsequent linear approximation of the gradient. He also recommends that random numbers be substituted in the function to observe the deviation of the response, and that the deviation be as an error in the characteristic value.

Both of these methods are capable of estimating the effect of internal noise factors. One distinguishing feature and, at the sametime, a disadvantage of these methods is the fact that error variation is observed only in nonlinear parameters. The effect of external noise factors is assessed based upon the deviation of the responses observed when external noise factors are allocated to an outer allocation table.

The Taguchi Methods recommend the "mixing of errors" between the "worst" sets of parameters that result in the maximum deviation in the positive direction (N1), and the maximum deviation in the negative direction (N2), so as to reduce the number of runs required. However, this should never be performed in simulations: although the direction of the maximum deviation coincides with that of the steepest gradient on the response surface, its erratic behavior prevents the same set of parameters from being generated every time.

In this case study, a response surface is created after substituting, all possible numbers between the upper and lower limits in the approximation equations, and then the variation in the characteristic value is measured. Shown in Figure 7 is the spatial representation of the experiment when three parameters are used. The advantages of this method include its higher efficiency, compared with the random numbering method and the assumption of non-zero values for the deviation, even at the limits on the response surface — a feature absent in the linear approximation. However, this method also has a disadvantage in terms that it does not necessarily include the treatment combination marked by the maximum sets of partial differential coefficients and the largest deviation.

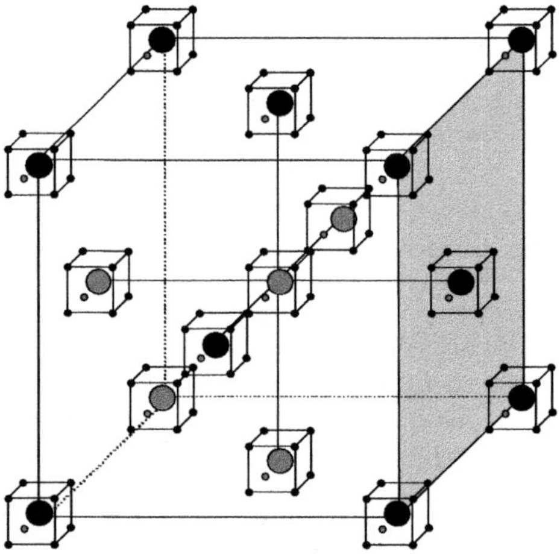

Fig. 7. Experiment Points for Error Estimation ($p = 3$, for example)

Assuming that dispersion occurs at a level of approximately 10% for each possible parametric value, a statistic (i.e. range) is calculated using the characteristic value prediction function of the optimization software. Shown in Figure 8 is the response surface generated by this calculation.

Lastly, multi-objective optimization is performed using two response surface functions. Multi-objective optimization is defined as the optimizing of multiple characteristic value responses simultaneously. Nakayama et al. [8] explain that the desirability index or the degree of deviation should first be computed using the target response and limits set for each characteristic value. The product is then multiplied by the individual weight of the value to determine the scalar. Accordingly, a single-objective optimization scheme is deployed to search for the optimal value for the scalar.

The approach used for the computation of the desirability and degree of deviation differs from one software to another. Here, a comparison is made between some software programs to detect differences in the approach: in the following experiment, 50 for the target value of the characteristic value and ± 20 for the limits (tolerance) are respectively specified.

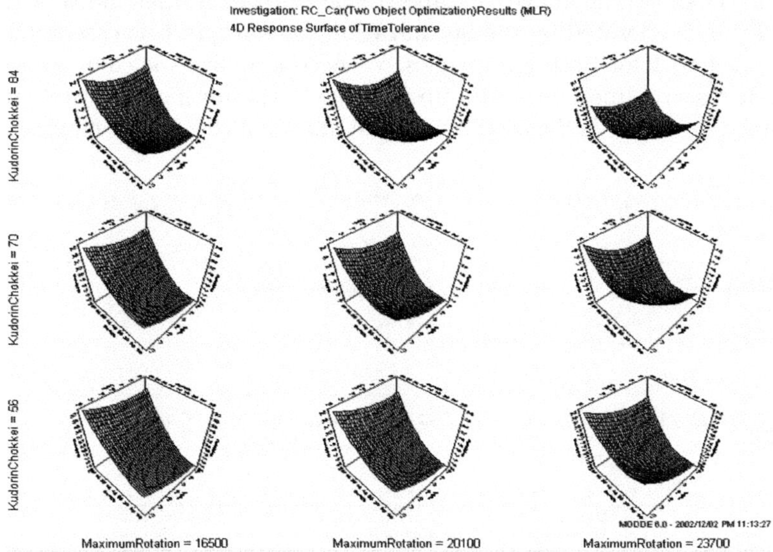

Fig. 8. Response Surface of Error (axis Right; TG–Left; GR, array Horizontal; MR–Vertical; KC)

JMP and JUSE-STATWORKS employ an approach introduced by Derringer and Suich [9], who defined a desirability function within the limits shown in Figure 9. Any response outside the constraints, irrespective of its distance from the limits, is assigned the number 0since it is regarded as less desirable.

MODDE [10] adopts another approach with a different function. As illustrated in Figure 10, any deviation with a response within the limits is assigned 0, whereas a deviation with a response outside the limits increases by the square of the distance from the limits. Whichever approach is selected, it is essential that users have a proper understanding of its characteristics.

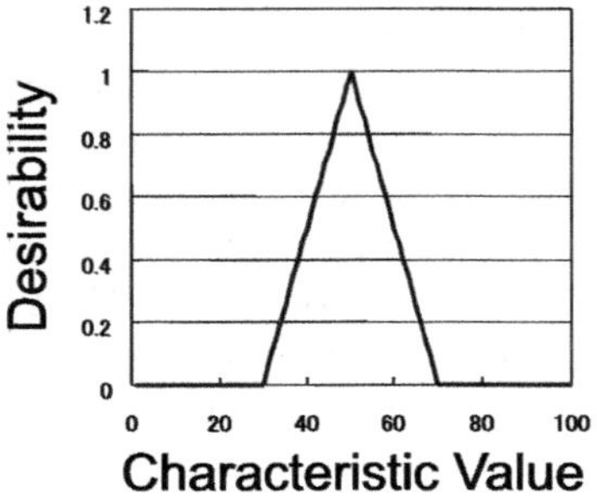

Fig. 9. Function of Desirability

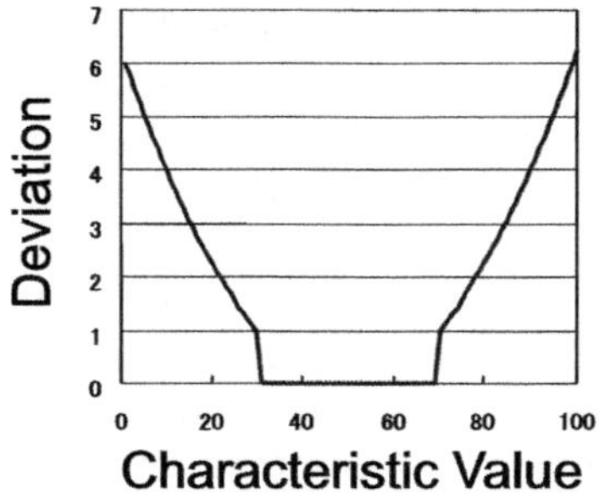

Fig. 10. Function of Deviation

Figure 11 shows the image that appears on the screen when multi-objective optimization is operated by the optimization software. In this case study, the weight is set to 1 for both approaches. When the analysis is commanded, an operation research method starts up and the optimum parametric setting is determined — which is denoted by the hatching pattern in the figure.

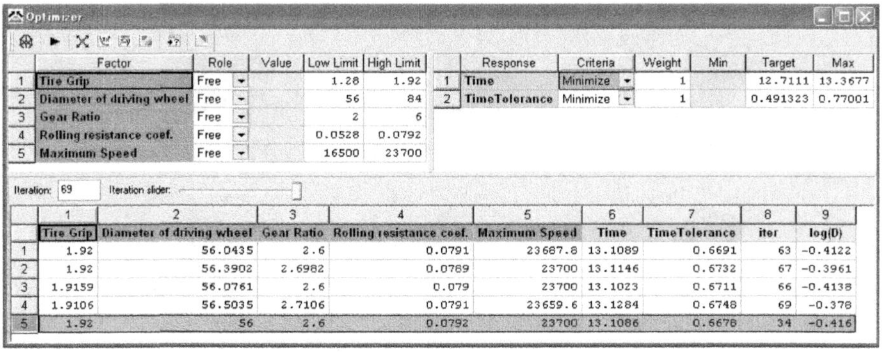

	Factor	Role	Value	Low Limit	High Limit
1	Tire Grip	Free		1.28	1.92
2	Diameter of driving wheel	Free		56	84
3	Gear Ratio	Free		2	6
4	Rolling resistance coef.	Free		0.0528	0.0792
5	Maximum Speed	Free		16500	23700

	Response	Criteria	Weight	Min	Target	Max
1	Time	Minimize	1		12.7111	13.3677
2	TimeTolerance	Minimize	1		0.491323	0.77001

	1	2	3	4	5	6	7	8	9
	Tire Grip	Diameter of driving wheel	Gear Ratio	Rolling resistance coef.	Maximum Speed	Time	TimeTolerance	iter	log(D)
1	1.92	56.0435	2.6	0.0791	23687.8	13.1089	0.6691	63	-0.4122
2	1.92	56.3902	2.6982	0.0789	23700	13.1146	0.6732	67	-0.3961
3	1.9159	56.0761	2.6	0.079	23700	13.1023	0.6711	66	-0.4138
4	1.9106	56.5035	2.7106	0.0791	23659.6	13.1284	0.6748	69	-0.378
5	1.92	56	2.6	0.0792	23700	13.1086	0.6678	34	-0.416

Fig. 11. Screen Image of Multi-Objective Optimization Software

5 Current Issues Surrounding Optimization Techniques

5.1 Background to Issues

The simplified parameter design processes of optimization software have resulted in the current situation where the majority of design engineers using the software are applying the SQC methods like a black box. The problem is that these design engineers might draw an erroneous conclusion as they lack basic knowledge required to interpret the simulation outputs and steps of the program. Especially when the outcome of such simulations is used to fit characteristic values, the concern becomes even greater. For the simulations differ from conventional SQC methods in many respects.

As mentioned earlier, simulation-based optimization technologies have developed and spread rapidly since around 1995, without any particular emphasis on educating and informing users. Thus, the problems are not attributable only to users. For example, Kashiwamoria et al. [11] used an orthogonal array to create a nonlinear response surface equation and performed ANOVA.

Some of the more major issues are exemplified and discussed in individual steps below. It is worth mentioning that there are also other problems too numerous to be comprehensively listed here.

5.2 Issues with Screening

ANOVA has a different meaning between simulations and actual experiments. Design engineers without this knowledge might draw a different conclusion of screening. In simulations, ANOVA will impart statistical significance to almost all parameters being measured if using conventional approaches.

5.3 Issues with Response Surface Modeling

RSM requires a design of experiments suitable for the objectives. Although optimization software proposes numerous suggestions, including D-Optimal, this might confuse design engineers who are not very skilled in the design of experiments. An orthogonal array is the Resolution III design, and thus an incorrect choice for RSM. However, it has been routinely applied in many studies solely because design engineers have grown accustomed to it.

Some design engineers have added experimental points near the external values of the response surface to attain convergence of approximations in order to "enhance the precision of the approximation solutions." However, thier estimation precision of the approximations is doubtful. These issues may remain undetected unless design engineers have sufficient expertise in design optimization and rotatability.

5.4 Optimization Issues

The major issue with multi-objective optimization is that an appropriate weighting method has not been established yet. Weighting must be properly redistributed to achieve the best solution without sacrificing some of the characteristic values, especially when there is an internal dependent relation between the responses. Further discussion is necessary to determine how to best calculate and weight error responses in robust optimization.

5.5 Issues Peculiar to Simulations

There are also more fundamental issues, which were left unstated in the case study section. Generally speaking, a response surface model is computed using the least square method, which assumes no deviation in measured responses and no approximation errors in the model (i.e. it is homoscedastic).

Simulations do not yield errors. To be more precise, they yield only a constant error with a one-sided deviation at each measuring point. This clearly violates the preconditions for the least-square method application, which has led to the argument that this method itself should be excluded from the analysis procedure. It would be best to replace it with the Penalty Method.

Also, estimating error responses in robust optimization while assuming the homogeneity of error is clearly self-contradictory. To remedy this condition, the design of experiments should be saturated at its onset, or higher-order equations should be set up to neglect all degrees of freedom when modeling the response surface. Some optimization case studies include a fourth- or fifth-order approximation equation, but that approach is utterly meaningless in the field of specific engineering technology.

Error modeling also has problems. An actual experiment is able to detect stochastic variation such as $1/f$ fluctuation and the experimental errors based on the resulting perturbation. In such a model, it is not necessary to take into consideration immeasurable or uncontrollable factors other than A, B, and C in Figure 6. On the other hand, a simulation has no means of estimating these factors. This exposes design engineers to a challenge of determining how to integrate these so-called noise factors into the model. A statistic cause and effect model describing the relationships between responses; parameters and noise factors will be the key to a solution. Further studies are highly anticipated.

6 Conclusion

In this paper, digital engineering technology was discussed with a special emphasis on optimization of design characteristics through the integration of CAE into SQC. The methods described herein have been widely used in the field of SQC for a long time. However, CAE-based optimization differs from actual experiments in the following ways:

A. CAE yields no error in repeatability. Therefore, repeatability of center-points in central composite designs and the degree of freedom hold no significance. Conventional statistic software users need to pay attention to this aspect.
B. An error in robust optimization is actually a deviation between responses, not an actual error.
C. Compared with the actual experiment-based Taguchi Methods, CAE-based optimization is unable to analyze robustness against true errors such as $1/f$ fluctuation. Hence, only the parametric responses that are transformed into nonlinear parameters are subject to estimation.

As mentioned in the foregoing, despite the rapidly growing popularity of these new methods, there are still a lot of critical issues to be resolved and publicized. These issues are summarized in Table 2. It is crucial that SQC researchers address these challenges and develop the existing SQC methods into more adequate forms.

7 Acknowledgements

We would like to express our utmost appreciation to the associates of Working Group One of the JSQC Research Conference on Simulations and SQC for their valuable opinions and advice on this paper.

Table 2. List of the current issues in CAE-SQC integration

Analysis Steps	Screening		Modeling (RSM in particular)		Optimization		
	Design of Experiments	Analysis of Results	Design of Experiments	Analysis of Results	Single-objective	Multi-objective	Multidisciplinary
Issues establishhed by statistics disciplines but not necessarily applied by other disciplines or commercial software.	· They are maybe cases where the Plackett-Burman and other factorial designs are used, which are more efficient than orthogonal arrays? · Factors are directly allocated to a large-scale allocation table to model a response surface without preliminary screening. · Some software has allocation problems.	· Due to the old way of handling the degree of freedom, it is not recognized that computer experiments are begin analized. · ANOVA is performed based on conventional approaches. (Statistical significance is not the same. Pooling effects of insignificant parameters makes ANOVA barely feasible due to deviations.)	· Orthogonal arrays are often used instead of CCD or other designs that are more suitable for determining the effect of interaction terms. · Allocation of interactions does not reflect the alia of relations within a design. · The increments between the levels of factors are not uniform. · All parameters are often allocated to inner allocation table.	· Widths between levels and weighting to orthogonalize linear terms against nonlinear are not adjusted. · Variable selection is not always performed. · Parameter perturbation is not always equalized at all points of the experiments in robust optimization.	· Some operation research-based software is designed for manual search for the optimum level.	· Two-stage programming design is used although parameter perturbation creates errors. · Enompassing index S/N ratio is used without simultaneously optimizing two characteristic value responses and dispersion. · No scalarization is carried out. (Continued above at right)	· In reality, optimization is executed in a series of experiments. · Simple multi-objective optimizations are sometimes referred to as multidiscipli-nary designs. · Simultaneous optimization is performed via higher-order response surface modeling without recognizing multimodal problems.
Controversial issues among statistics disciplines.	· Whether external noise factors can be allocated to an inner allocation. (Can they be screened within an *inner allocation* table even though interactions between internal and external noise factors are relevant.) · What is permissible level of confounding. · How to design experiment for parameter set including specific factors with known interactions.	· How to extract significant factors when ANOVA is not used. How to detect nonlinear (nonlinear *or higher-order*) characteristics in two-level factorial designs. (Adding Centerpoints)	· How to handle points of repeated experiments where simulations only supply an identical value. · *How to handle the* degree of freedom if the experiment cannot be saturated. · Whether numerals assigned to screened-out variables should be fixed or perturbed?	· Whether variables in multi-objective optimization should be selected jointly or separately according to each *objective function.* · Which is adequate range of deviation at each measuring point in robust optimization. Relationship with weights. · Whether computing dispersion with linear-approached second moment method is appropriate.	· Whether robustness should be defined as point of minimum dispersion or point that satisfies all constraints *including those for* variance. · How problematic it is to add a provisional limit as point of experiment and create a new model to obtain convergence.	· What is best among several methods available to integrate desirability or deviation of *characteristic value* responce in single index based on optimum for subsequent scalarization. · How to reconcile disagreement on how to weigh the differences between objective functions. · How to handle the characteristics of tradeoffs.	· When optimization is performed in a series of experiments, experimental procedures are not unified. (MIT's DSM method is one of best known.)

Unresolved issues.	· How to apply supersaturated design. · How to apply a uniform design to analyze characteristics with multimoda functions.	· Whether any parameter for robust optimization is ever excluded at the screening stage. (i.e. Parameters with error responses transformed to fit nonlinear characteristic values always influence characteristic values and thus are always statistically significant at screening.)	· What is optimal allocation method to outer allocation table for external noise analysis (Orthogonal or Monte Carlo method). · In robust optimization, optimization, what is optimal scale ofdeviation to be given to parameter at each measuring. (Smaller than in best-subset selection prosedure) · In the Monte Carlo method, which is optimal scale of deviation. (Smaller than in best-subset selection procedure) · How to apply a uniform design to analyze multimodal characteristics.	· When calculating dispersion from range of deviation, whether characteristic value should be calculated based on approximation equation or another analysis through CAE. · How to select statistic measuring characteristic value responses when using response surface model of variance. · Whether least-square method computes correctly penalty model of good fit.	· What are rules for adding points of experiment to enhance accuracy of response surface model (Where they should be added). · What is a design that allows such addition while saving previous points of experiment in memory.	· How to optimize parametric settings when there is internal dependent relation between objective functions. · How to weight dispersion. · How to determine the range of limits for dispersion. · How to estimate effect of factor not successfully parametrized in simulations. (Whether a cause and effect model is available.)	· Methodologies have not been established yet.

References

1. Oda, Z. (The Japan Society of Mechanical Engineers, ed.) (1989) Optimum Design of Structural Materials, Gihodo Shuppan, 246.
2. Kawamo, K., Yokoyama, M. and Hasegawa, H. (2000) Foundation and Application of Optimization Theory, Corona, 186–194.
3. Kashiwamura, T., Shiratori, M. and Yu, Q. (1997) Evaluation of Dispersion and Structural Reliability in the Statistical Optimization Method, Transactions of JSME A, 63. [610]. 1348–1353.
4. Yamakawa, H. (2000) Trends of Multidisciplinary Design Optimization, JSME OPTIS2000 Special lecture
5. http://member.nifty.ne.jp/QYR03001/calc/calctop.htm
6. Myers, R.H. and Montgomery, D.C. (2002) Response Surface Methodology: Process and Product Optimization Using Designed Experiment 2nd edition, John Wiley & Sons Inc., USA, 557–586.
7. Akagi, S. (1991) Design Engneering (2), Corona, 112.
8. Nakayama, H. and Yano, T. (The Society of Instrument and Control Engineers, ed.) (1994) Multipurpose Planning Theory and Application, Corona, 38.
9. Derringer, G. and Suich, R. (1980) Simultaneous Optimization of Several Response Variables, Journal of Quality Technology, 12. [4]. 214–219.
10. MODDE Online manual
11. Kashiwamura, T., Shiratori, M. and Yu, Q. (1998) Optimization of Non-Linear Problem by DOE, Asakura Publishing Co., Ltd.

Application of DOE to Computer-Aided Engineering

Ken Nishina[1] and Mutsumi Yoshino[2]

[1] Nagoya Institute of Technology
nishina@nitech.ac.jp
[2] DENSO Corporation
yoshino@prd.denso.co.jp

Summary. The simulation experiment by Computer-Aided Engineering (CAE) has been developing instead of the actual experiment and it is surely one of the powerful supports to the concurrent engineering. In the actual CAE working, however, some issues on the simulation results appear. How to cope with them often depends on some experimental or institutional skill of the CAE engineers. It may be an obstacle to progress of the reduction of lead-time. In this article, we indicate some serious issues on CAE working. Especially, we take up calibration to recover inconsistency of the numerical solutions by simulation experiment with the results of actual experiment. We propose how to cope with calibration by using DOE.

Key words: Simulation experiment, Actual experiment, Calibration

1 Introduction

The recent advances in Digital Engineering have been contributing the reduction of lead-time of the product development in business process. Especially in the product design and the production process design, the simulation experiment by Computer-Aided Engineering (CAE) has been developing instead of the actual experiment and it is surely one of the powerful supports to the concurrent engineering. In the actual CAE working, however, some issues on the simulation results appear. How to cope with them often depends on some experimental or institutional skill of the CAE engineers. We cannot deny that it can be an obstacle to progress of the reduction of lead-time.

A proper introduction of the Design of Experiment (DOE) to a series of working in CAE is needed in order to improve the working efficiency in CAE. It need scarcely be said that DOE has been applied to industry such as Taguchi Method. But almost of them has been done to actual experiments. We can get some controversial point issues in the case of application to simulation experiments.

On the above background, the Japanese Society for Quality Control (JSQC) established a research group, which of the title is "Simulation and SQC". This article presents an outcome of a working group (WG1) of the research group. The main mission of the WG1 is to indicate some serious issues in CAE working and propose how to cope with them by using DOE.

From the discussion in WG1, we try to propose a proper application of DOE to improve the working efficiency in CAE. Our controversial point issues are as follows:

1. Consistency of the numerical solutions with the results of actual experiment
2. Selection of the simulation model and data analysis for the robust optimization

This article presents some discussions about the above issues in WG1. However, the main content is the first issue. Concerning the second issue, some essential points are only introduced. In addition, some remaining issues are indicated.

CAE assumes that the result of a simulation precisely reproduces the result from actual experiment. The first point above means that this assumption does not hold. To achieve the reproducibility, a simulation model needs to be calibrated "forcibly". "Forcible calibration" means that some parameters are altered from the theoretical value and some constraint conditions are changed to meet the numerical solutions by simulation experiment with the results of actual experiment for some specific points. This is referred to as "calibration". In this article, the calibration is explained through an example of a cold forging.

2 Example of calibration in CAE working

2.1 Outline of the example

The subject of this example is the pin-like protrusion manufactured by cold forging as shown in Fig. 1, which is a part of vehicle seat. The effect of the part is height shown in Fig. 2. The design parameters to be optimized are the punch depth (d), the punch width (b) and the shape of the chamfer of the die (w, h) as shown in Fig. 2.

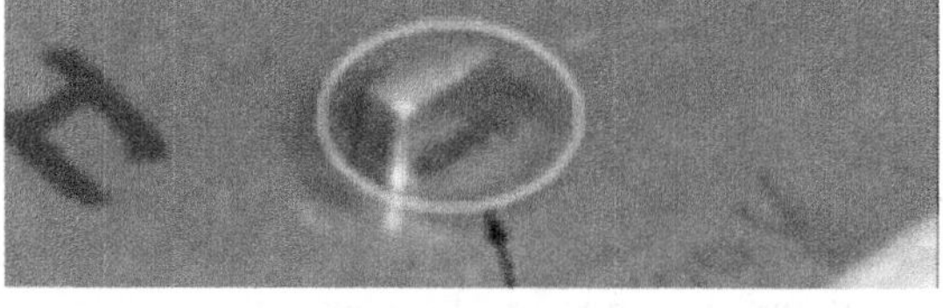

Fig. 1. Pin-like protrusion (presented by N. Taomoto)

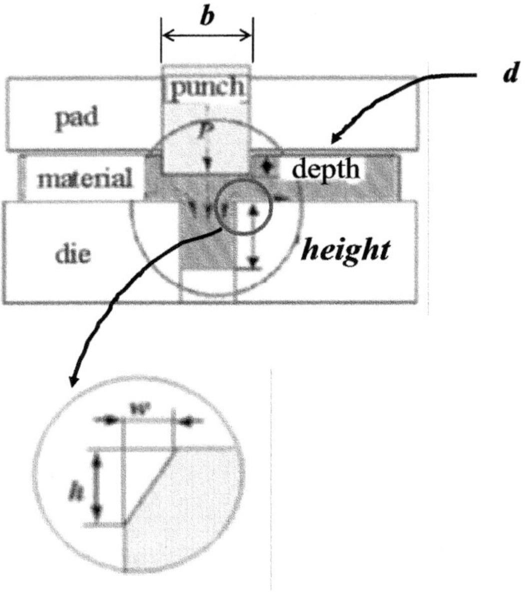

Fig. 2. Design parameters (punch depth (d), the punch width (b) and the shape of the chamfer of the die (w, h)) and effect (height) (presented by N. Taomoto)

2.2 Calibration in the example

The actual experiments are so inefficient because any change in the shape of the chamfer would require the fabrication of a new die. Therefore, simulation experiment was carried out.

Prior to optimization, the actual experiment and simulation experiment were performed only on the existing die. That is, in the actual experiments protrusion height is obtained with varying the punch depth, and the other design parameters are assigned as a constant level considering the facility of the actual experiment. Fig. 3 shows the comparison between the results of the simulation experiment and of the actual experiment. Then the comparison shown in Fig. 3 did not reveal reproducibility between the results of the simulation and the actual experiment.

After trial and error, the CAE engineer was able to bridge the non-reproducibility by changing the constraint conditions of the friction between the die and the material.

2.3 Controversial points from the example

From this example, we can indicate three essential controversial points on the calibration as follows:

1. In the above actual experiment, only the punch depth, of which the level can be easily changed, is varied.

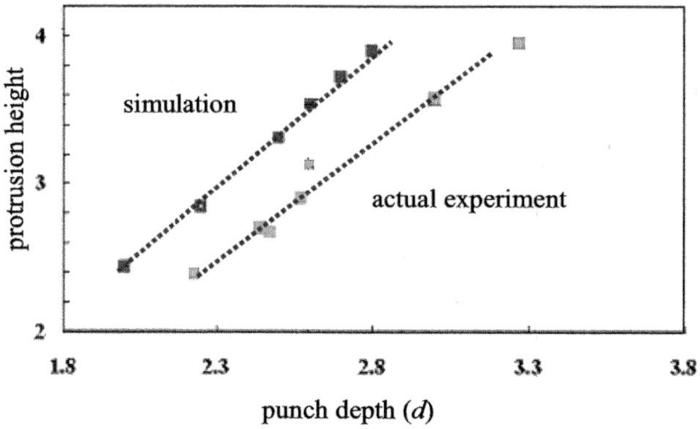

Fig. 3. Comparison of the results of the actual experiment and of the simulation experiment (presented by N. Taomoto)

2. There is no criterion for the calibration.
3. After trial and error, the CAE engineer was able to bridge the non-reproducibility.

As explained earlier, some design parameters keep constant in the actual experiment for the calibration. Our controversial point is the reproducibility of the actual experiment in the case of the optimization by the calibrated simulation model.

Over-calibration owing to the over-fitting to the results of the actual experiment may introduce the non-reproducibility. A criterion for the degree of calibration should be needed.

The last point is efficiency of the calibration. The efficiency in the procedures of the calibration considerably depends on the experimental and institutional skill of CAE engineer. So we recommend the application of DOE to the calibration instead of trial and error.

The application of DOE has been developed in some softwares. However, they aim to only the calibration to the results of the actual experiment; on the other hand, our proposal is that the first and second controversial points above are considered.

3 Proposal for efficient calibration

3.1 Factors of DOE in calibration

The parameters of the simulation model which may be used for the calibration are referred to as "calibration factor". In general the materials property index such as Young's modulus, Poisson's ratio, density and yielding point are designed as the calibration factors, which are the variable factors. In addition, whether or not to set the constraint conditions (translational constraint and rotational constraint) are designed as the calibration factor, also. These are the attribute factors. In the example presented in Sec. 2, the calibration is carried out by removing the constraint conditions of the friction between the die and the material.

In the example in Sec. 2, the design parameters are the shape of the chamfer of the die and the punch depth. Then the punch depth was selected as a factor in the actual experiment. Such a factor is referred to as "signal factor", which should be selected out of the design parameters from the viewpoint of being easy in carrying out the actual experiment. The number of the signal factor should be small because of decreasing the number of the actual experiment.

3.2 Proposal for efficient calibration

3.2.1 Calibration factors and signal factor

At first, the calibration factors are selected. As mentioned earlier, the constraint conditions (translational constraint and rotational constraint) are designed as the attribute calibration factors. These factors are assigned to the two levels, that is, "free" and "constraint". The materials property index such as Young's modulus, Poisson's ratio, density and yielding point are designed as the variable calibration factors. These factors are assigned to the three levels. The theoretical value of the material property is assigned to the second level. The first and third levels are assigned to the theoretical value ± about 10%, respectively.

As mentioned earlier, a signal factor is selected out of the design parameters from the viewpoint of being easy in carrying out the actual experiment. The signal factor is assigned to more than or equal to three levels. If the number of the level is three, then the level expected to be the optimal condition should be assigned to the second level.

Fig. 4 shows an example of the design of experiment for the calibration in order to identify some active calibration factors. As shown in Fig. 4, the orthogonal array is useful (Fig. 4 shows L_{18} as an example). The calibration factors are assigned to the inner factors and the signal factor (G) is assigned to the outer factor.

The actual experiment is carried out only with the signal factor and at least two repetitions. As mentioned in Sec. 2, in the actual experiment the other design parameters are assigned as a constant level considering the facility of the actual experiment.

If the number of the calibration factors is too large, then the experimental design correspondent to the inner orthogonal array shown in Fig. 4 should be divided into two orthogonal arrays. One is designed for the attribute calibration factors. The other is for the variable calibration factors. A two-level orthogonal array is applied to the former case, and a mixed-level orthogonal array is done to the latter case.

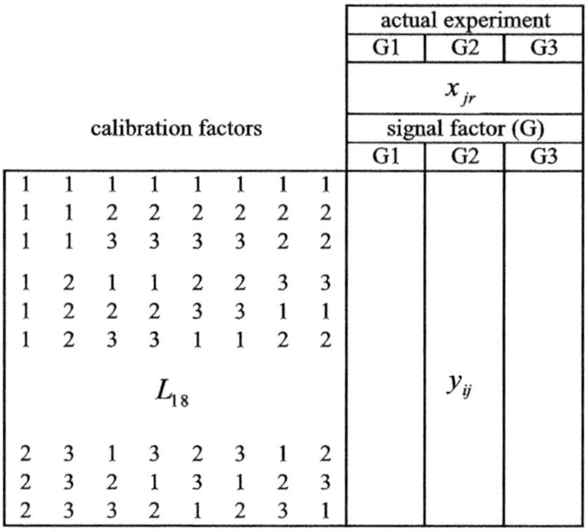

Fig. 4. An example of the experiment for the calibration

3.2.2 Data Analysis

Let x_{jr} be a result of the actual experiment at G_j of the signal factor and at the rth repetition (see Fig. 4); and c_{ij} a numerical solution by the simulation experiment at the ith experiment of the inner orthogonal array at G_j. Then y_{ij} denotes the difference between the actual experiment and the simulation experiment as follows:

$$y_{ij} = c_{ij} - \bar{x}_{j\cdot}, \quad \bar{x}_{j\cdot} = \frac{\sum_{r=1}^{R} x_{jr}}{R} \qquad (R\text{: the number of the repetition}).$$

The difference y_{ij} means "the nominal (zero)–the best" character, which comes from the Taguchi method. So we propose applying the Taguchi's two step method, given for example by Taguchi (1993).

The first step is that the condition such that the variance (see Equation (3.1)) is minimized is determined. The analysis is referred to as the calibration for stabilization. Then the analytic character should be as follows:

$$S_i = \log\left(\sum_{j=1}^{J} (y_{ij} - \bar{y}_{i\cdot})^2\right), \tag{3.1}$$

$$\text{where } \bar{y}_{i\cdot} = \frac{\sum_{j=1}^{J} y_{ij}}{J} \qquad (J\text{: the number of level of the signal factor}).$$

The next step is that the bias from the nominal value (zero) is calibrated. The analysis is referred to as the calibration for trueness. Then the analytic character should be $\bar{y}_{i\cdot}$. The conditions such that $\bar{y}_{i\cdot}$ is equal to zero may be countless; however, the priority of determining the calibrated condition should be given to the determination at the first step.

3.2.3 Verification of the calibration

Let c_{0j} a numerical solution by the simulation experiment with the calibrated condition at G_j. Then the ratio of the following two statistics, F, should be given as the evaluation index for the calibration:

$$F = \frac{\sum_{j,r}(x_{jr} - c_{0j})^2}{JR} \Bigg/ \frac{\sum_{j,r}(x_{jr} - \bar{x}_{j\cdot})^2}{J(R-1)}. \tag{3.2}$$

The numerator and the denominator of Equation (3.2) mean the luck of calibration and the experimental error of the actual experiment, respectively. The statistic F can be applied in the framework of the statistical test.

3.3 Discussion

As mentioned in Sec. 2.3, we have indicated the three essential controversial points on the calibration. We can give a solution for the two points of them (no criterion for the calibration and inefficiency for calibration (trial and error)) by applying the DOE as proposed in Sec. 3.1 and 3.2.

The remaining controversial point is that the actual experiment is carried out on the condition that the design parameters except the signal factor are assigned as a constant level. Our concern is the reproducibility of calibration when the other design parameters are varied; however, we have no way to confirm the reproducibility. Our proposal against the reproducibility is the calibration for stabilization, that is, the first step of the data analysis. Of course, we can not say that it assures the reproducibility. But the calibration for stabilization means that the reproducibility is realized at least within the signal factor.

4 Other discussions on the optimization after the calibration

After the simulation model is identified by the calibration, it is supposed that the robust optimization is performed by using an approximated second-order polynomial equation model. Then, a large number of the design parameters are an obstacle to the optimization because a so large scale DOE for the robust optimization have to be conducted. So the design parameters should be screened.

In general, a Resolution III two-level design of experiments is employed in screening the design parameters; however, center-point level runs are often added.

There are three issues in the usual method above as follows:

1. The significant test can not be carried out.
2. The interactions are not considered.
3. The large number of runs is needed when the number of the design parameter is very large.

Yoshino *et al.* (2006) optimized the design parameters of a radio control car to minimize its rap time, which can be downloaded from a website. At that time, the amount of effects is expressed by using the Pareto diagram. Judging from the Pareto's law, five design parameters from the sixteen design parameters were selected. Instead of the significant test, the stand point from the descriptive statistics as Yoshino *et al.* (2006) is applicable.

Our concern related to the robust optimization is that the interactions are not considered in screening the design parameters. In the usual method the design parameters are screened out judging from the amount of the first order effect. It is supposed that if a design parameter (x_1) is screened out because its linear effect is small, although the interaction between the design parameter and the other (x_2) with the large linear effect is not so small. Then the terms x_1, x_1^2 and x_1x_2 are omitted in the second-order polynomial equation model.

The design parameters selected in the simulation model are perturbed when the robust optimization. The lack of the term x_1x_2 produces an effect not so small on the robust optimization all the greater because the main effect of x_2 is large. Therefore, a Resolution IV design of experiments could be employed in screening the design parameters.

The third issue is that the number of runs is too large in screening the design parameters. This is a contradiction with the discussion above. Application of the supersaturated design is considered as a countermeasure although orthogonality is sacrificed a little (see Yamada (2004)).

5 Conclusive remarks

We have discussed the practical issues and given a proposal on the optimization problem using the simulation experiments by the CAE. As mentioned earlier, the content is an outcome of the WG1 of the research group "Simulation and QC". The WG1 is now in action and our mission is unchanging.

We are posing some remaining issues as follows:

1. Verification of the proposed method for the calibration through some case studies
2. Calibration in the case of the multi-characteristic

3. Resolution peculiar to the robust optimization using the simulation experiment
4. Building an approximation model for complicated physical phenomenon such as multi-modal pattern

As is generally known, DOE originally came from the field of agriculture, and then transferred to the field of engineering. However, DOE have been mainly developed in the actual experiments. We recognize that some practical issues on application of DOE to CAE have not been resolved.

It is well known that the "design quality" is evaluated as the reproducibility in the market. It should be noted that in using the simulation experiment, the first step toward the reproducibility is to build the simulation model, which can reproduce the actual experiment.

Acknowledgement. The authors would like to thank the members of the WG1 of the research group "Simulation and SQC" of the Japanese Society for Quality Control for their useful comments.

References

1. Taguchi, G. (1993). *TAGUCHI on Robust Technology Development*, ASME Press.
2. Yoshino, M. and Nishina, K. (2006) SQC and Digital Engineering — Technological Trends in Design Parameter Optimization and Current Issues —, *Grammar of Technology Development*, 129–145.
3. http://member.nifty.ne.jp/QYR03001/calc/calctop.htm
4. Yamada, S. (2004) Selection of active factors by stepwise regression in the data analysis of supersaturated design, *Quality Engineering*, 16, 501–513.

Part III

Statistical Methods for Technology Development

A Hybrid Approach for Performance Evaluation of Web-Legacy Client/Server Systems

Naoki Makimoto[1] and Hiroyuki Sakata[2]

[1] University of Tsukuba
makimoto@gssm.otsuka.tsukuba.ac.jp
[2] NTT Data Corporation
sakatahr@nttdata.co.jp

Summary. In this paper, we propose a hybrid approach to evaluate the performance of Web-legacy client/server systems. The approach combines three methods, hardware and software simulations and approximate analysis. Hardware simulation measures basic performance characteristics of the system whose data are utilised in the subsequent methods for designing the system so as to satisfy the prescribed quality of service. From our experience of its application to a prototype system, we observed that both software simulation and approximate analysis provide reasonable evaluation of the system performance.

Key words: simulation, Web-legacy C/S system, performance evaluation, quality of service

1 Introduction

According to rapid growth of the Internet, a number of companies have started to develop a Web client/server (C/S) system as a part of their business process. Internet shopping and banking are typical services provided by such systems. On the other hand, as the Internet businesses prevail among our society, an increasing number of accidents caused by insufficient capacity of the system are reported. Adequate capacity planning of the system therefore plays a key role to manage such business process successfully. Lack of capacity could cause a serious congestion and instability of the system while too much capacity leads to high cost of investment.

When developing such Web C/S systems, it is often the case that the system contains both Web client and legacy systems. These two systems have quite different characteristics since Web client server is a rather new technology while the legacy system has a long history. In addition, statistical properties of arriving requests through the Internet differ from those of the legacy system. Therefore, existing techniques for performance evaluation and capacity planning are not directly applicable to such system.

In this paper, we introduce a hybrid approach applied to evaluate the performance of Web-legacy C/S systems. The term *hybrid* means that we combine two types of simulations, hardware simulation and software simulation, together with approximate analysis. We applied this approach to evaluate the performance of a prototype system. From our experiments, both software simulation and approximate analysis are shown to provide reasonable estimations of the measurements obtained by hardware simulation for systems with relatively simple structure.

This paper is organized as follows. In the next section, we describe a Web-legacy C/S system under consideration. Hardware simulation is then explained in some detail. According to the traffic patterns, experiments are classified into 3 types: single-profile, multiple-profile and random access. Based on the specification of the system, we next construct a logical network model for software simulation. Parameters of the model are estimated from the measurements of the hardware simulation. From the results of software simulations, we identify a queueing model for the most congested part of the system. A simple approximation formula is available from queueing theory which gives a rough estimation of the turn around time. In Sect. 4, experimental results are provided to check the usefulness of the approach. Finally in Sect. 5, we conclude this paper by stating some remarks.

2 A Web-legacy C/S system

In this section, we will describe the system whose performance evaluation is the main issue in this paper. The system is called Web-legacy C/S system since it combines advanced Web technology with the existing legacy system as explained below.

Fig. 1 depicts the hardware configuration of the system. At the front-end of the system, there is a cluster of World Wide Web (WWW) servers that manages connection between many types of clients (cellular, PC, mobile, PDA, etc.) with the system. At the back-end of the cluster of WWW servers, there is a cluster of application servers that administers the main business logic program and connects to the legacy system allocated at the back-most side. The legacy system was set up in relatively early days and still provides such services as online-banking and scientific computation to limited users. In addition, there is a legacy access server as a gateway between the application servers and the legacy system.

The main features of the system can be summarised as follows:

- Each server adopts *multi threaded server architecture* that enables parallel processing. *Thread* is a primitive resource for all types of requests to be

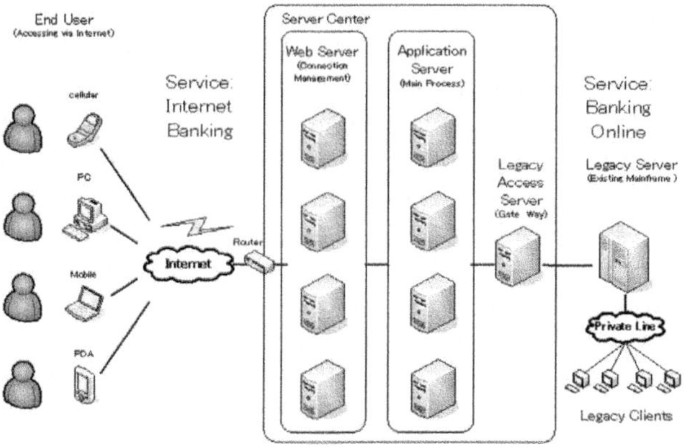

Fig. 1. A Web-legacy C/S system

processed. Specifically, an arriving request directly goes into its own service if there is a vacant thread, otherwise it must wait at a queue until a thread becomes available.

- Since the legacy server was set up long before than the front Web system, it is sometimes difficult to estimate its performance measures such as response time and throughput, when interconnected with the Web system.

When developing this type of system, it is necessary to satisfy a prescribed quality of service (QoS). Typical examples of QoS are such that, the mean response time must be less than 4 seconds, and at least 90% of requests must be responded in 10 seconds. Among other design parameters, number of thread is a very important parameter of the system since it determines load to the legacy system whose performance is difficult to predict as stated above.

Before closing this section, we briefly explain the prototype system developed for this research to check effectiveness of the approach. In the prototype system, each server is single configured (i.e., single-machine, multi-threaded). Also, the legacy system is substituted for *stub process (program)* that emulates processing delay of the legacy system. The delay distribution can be chosen among specific distributions such as exponential and log-normal, so as to fit the real data.

3 A hybrid approach for performance evaluation

In this section, we explain a hybrid approach applied to evaluate the performance of a Web-legacy C/S system described in Sect. 2. The term *hybrid* means that we combine two types of simulations, hardware simulation and software simulation, together with an approximation analysis. First, we give an outlines of our approach. Details are explained in the following subsections.

- Hardware simulation
  A traffic emulator is used to generate an artificial stream of arriving requests which are processed by the system under consideration. We measure such performance as turn around time (total time spent in the system), execution time of each sub-process, and ratio of unfinished requests.
  The main purpose of the hardware simulation consists of two folds. The first one is to collect data of processing times of all sub-processes of the system. The data will be used to estimate input parameters in the subsequent software simulation and approximate analysis. The second purpose is to capture the performance characteristics of the system. Because the whole system contains sub-systems supplied by third parties, combinations of these sub-systems could give an impact on the performance to some extent.
  Among three methods, the hardware simulation provides the most accurate results. On the other hand, the hardware simulation is the most expensive and time-consuming. As a result, it is difficult to collect data by the hardware simulation for various sets of parameters.
  In addition to the knowledge of the real system, hardware simulation requires skills for implementation and knowledge of simulation.
- Software simulation
  According to the specification of the system, we construct a logical simulation model for software simulation which approximately describes the behavior of the real system. Parameters of the model such as execution time of each sub-process are estimated base on the measurements of the hardware simulation.
  The main purpose of the software simulation is to evaluate the performance of the system for various sets of parameters. Since software simulation can be executed much faster than the hardware simulation, we can evaluate the performance by changing the input, capacity and configuration of the system. These results are used to decide appropriate design of the system so as to satisfy the required quality of service.
  Accuracy of the software simulation depend on accuracy of the logical model we construct and accuracy of the estimated parameters. Since all the detailed mechanisms of the real system can not be taken into account in the logical model, it is important to extract elements which give a large impact on the performance.
  Similarly to the hardware simulation, appropriate implementation of software simulation requires skills and knowledge of the software and simulation.
- Approximate analysis
  For a distributed system, it is often the case that the performance of the whole system is mostly determined by a bottleneck. This fact indicates the possibility to develop a simple approximate analysis by focusing on the bottleneck. The objective of the approximate analysis is to construct

a queueing model of the bottleneck and derive formulas for performance measures by invoking queueing theory.
The results provided by this method is in principle less accurate than the other methods. The main advantage of the approximate analysis is in its simplicity. In contrast to the simulations methods, the approximate analysis can be used without any specific knowledge since it is enough to plug a few parameters into formulas. Thus, these formulas are useful as a quick reference for engineers and practitioners.

Table 1 summarises the features of the three methods described above.

Table 1. Trade-off between three approaches

	accuracy	cost	implementation	main users
hardware simulation	accurate	high	difficult	researcher/engineer
software simulation	medium	medium	medium	engineer/researcher
approximate analysis	rough	low	easy	practitioner/engineers

3.1 Hardware simulation

In the hardware simulation, we use a traffic emulator to generate an artificial stream of arriving requests. These requests are executed by the real system and then returned to the emulator which measures the turn around time. We also measure waiting times and execution times of all primary sub-processes.

Since the hardware simulation is expensive and time-consuming, it is requested to collect data as effectively as possible. To this end, we design our experiment in the following three steps.

1. Single profile method
   In a single profile method, the traffic emulator generates a request after the execution of the preceding request has finished. Thus, only a single request is executed at a time in the whole system. The objective of this single profile method is to collect data of execution times of all major sub-processes which will be used in the subsequent software simulation and approximate analysis. Kino [4] emphasises an importance of the single profile method as a building block for performance evaluation. In our experiment, we focus on 10 primary sub-processes such as output log, check format, create session, and so on.
2. Multiple profile method
   In a multiple profile method, the traffic emulator generates some fixed number of requests concurrently. The system will process some but not all of the requests simultaneously and return them to the emulator after completing execution.

A typical outcome observed in this experiment is that a fixed number of requests are returned periodically with a fixed interval. This is a natural phenomena because all the requests are in effect waiting at the bottleneck queue. More specifically, the same number of requests as the number of processors at the bottleneck queue are returned at the same time. Also, the fixed interval of the periodical return is the same as that of the processing time at the bottleneck.
We call this procedure to identify a bottleneck as *localisation* which plays a role in the subsequent approximate analysis.
3. Poisson access method
Poisson access means that the emulator generates requests according to a Poisson process. A Poisson process is a mathematical model of random occurrence of events and widely used as an arrival process of requests for telecommunication theory. The main purpose of the Poisson access method is to collect data under the most realistic circumstances.

To give a concrete description of the hardware simulation, we will show the result obtained by applying the method to the prototype system. The details of the experiment are as follows.

- Traffic emulator
As a traffic emulator, we used LoadRunner Ver.7.0.2 produced by Mercury Interactive Corporation (http://www.mercury.com/). LoadRunner is a traffic generator that can emulate many concurrent users who put HTTP requests into the system according to given parameters (user-think time, arrival rate/distributions, repeat count, etc.).
- Configuration of the prototype system
The prototype system consists of a Windows 2000 server on which LoadRunner runs, two Solaris (Sun-O/S) servers for WWW-server/application-server/legacy-access-server and legacy stub application.
- Parameters
A list of important parameters used in the experiments are as follows.
  - number clients ranges from 1 to 20.
  - arrival rate of requests: 1.0, 1.5 and 2.0 (corresponding traffic intensities are 0.4, 0.6, 0.8, respectively).
  - number of threads: connection 3, main 20 and legacy 5
  - size of all buffers are set to 65535 which assures no blocking.
  - execution time in the legacy system is fixed as 2000 msec.

Under the above circumstances, we measured the turn around time and execution times of 10 sub-processes in the system. The execution time of three primary sub-processes measured in the single profile method are shown in Table 2. Since the execution time in the legacy system is fixed to 2000 msec. in this simulation, the turn around time is about 2014 msec.

The multiple profile method was executed by increasing the number of concurrently arriving clients from 2 to 20. As can be expected from the data

Table 2. Execution time of three sub-processes in the single profile method

sub-process	connection	main	legacy
exec. time (msec.)	5.39	5.56	3.05

from the single profile method and the number of threads, the legacy system is a bottleneck of the system. Specifically, when 20 requests arrive concurrently at the system, then 5 requests are replied almost concurrently every 2000 msec., meaning that 5 requests are served simultaneously in the legacy system. It can be also estimated that the legacy system operates like a multiple server system rather than processor sharing since the service rate does not deteriorate as the number of busy servers increases.

The results of the Poisson access will be reported in Sect. 4 to compare with the results obtained by other approaches.

3.2 Software simulation

Roughly speaking, the following three steps are necessary for software simulation: (1) construct a logical model of the system, (2) parameter estimation, (3) programming and implementation.

In the first step of the software simulation, we construct a logical model which emulates the behavior of the real system. The logical model is typically described by a queueing network consisting of several local queues through which a request proceeds. A local queue is composed of service places and waiting spaces. A request entering each local queue has to wait if it finds all service places are occupied, otherwise a service will start immediately. Likewise a processor sharing system, several service places can share more than one processors. Even service places in different local queues can share a processor, depending on the configuration of the system.

Control mechanisms of the network should be also took into account because they could give a strong impact on the performance. Examples of the control mechanisms include acceptance/rejection control, routing control, connection setup, sojourn time limit, garbage collection and others. To implement these control mechanisms, we may need to introduce additional transactions and queues into the logical model.

Parameters of the simulation model are estimated based on the data from the hardware simulation. Specifically, service times of local queues can be estimated by the data from the single profile method. Data from the multiple profile method could be used to see whether or not a processor is shared by some local queues. If a processor is shared, then the response time must increase as the number of simultaneously arriving customers increases while the response time is robust when local queues has multiple servers.

Once the logical model is constructed and its parameters are at hand, we will implement the simulation and check the model by using the data from

Poisson access. The software simulation is satisfactory if the results of software simulation are close enough to the data of the hardware simulation and exhibit similar tendency for different sets of parameters. Otherwise, we need to modify either logical model or parameter estimation by carefully investigating the reason of the deviation between them.

The above steps of the software simulation for the prototype system under consideration can be described as follows.

1. Construction of a logical model
   We construct a queueing model depicted in Fig. 2 for the prototype. The model consists of four local queues, a) connection, b) main, c) legacy access, d) output stream buffer. The legacy system is not modelled as a local queue in the logical model. Instead, it is considered as a server which exogenously determines the execution time at c). An arriving request goes through the network in the order of a), b), c) and d).
2. Control mechanism
   In the model, we introduce two control mechanisms, namely, setup connection with legacy system and garbage collection:
   - to setup a new connection with the legacy system at c), the legacy access server should open physical (device-level) and logical (protocol-level) connection. Generally, this setup process consume several overhead time.
   - execution of all the transactions in the system are periodically suspended for a certain period of time due to the garbage collection. This mechanism is modelled in the simulation since it is expected to degrade the performance.
3. Parameter estimation
   Execution time of the four local queues a), b), c) and d) are estimated

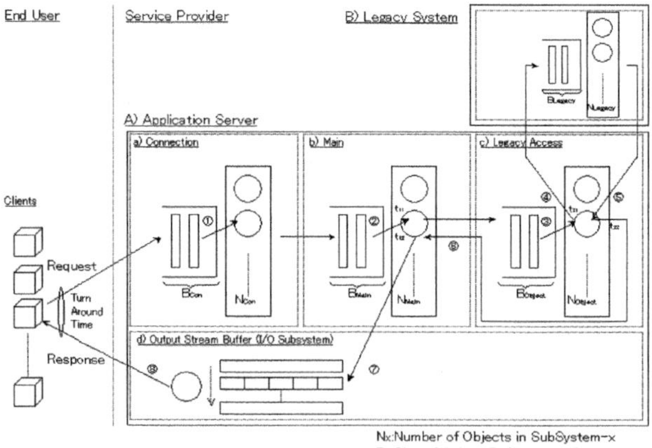

Fig. 2. A logical model for the software simulation of the prototype system

from the data of the single profile method. The interval of the garbage collection are set to their average measured in the Poisson access of the hardware simulation (the actual length and interval fluctuate depending on the state of the system).

We implemented the software simulation by SLAM ver.3.0 produced by Symix Systems Inc., Pristker Division. Comparative results for Poisson arrival of requests will be shown in Sect. 4. Overall, the results fit the data of the hardware simulation for this prototype system. It should be remarked here that execution times for implementing software simulation are just 2 to 3 minutes to get results accurate enough from practical viewpoint. Therefore, it is easy to evaluate the performance by changing, for example, the number of threads to find out the optimal value of this important design parameter.

3.3 Approximation analysis

In this step, we construct a queueing model by focusing on the bottleneck of the system which we already found by the localisation based on the multiple profile method of the hardware simulation. Concerning the objective of the approximate analysis to derive a formula for performance measures of interest, the queueing model should be as simple as possible. Possible candidates for queueing models include single stage queues such as M/G/c, G/M/c, GI/G/c and queueing networks which have so-called product form solution [1].

For the prototype system under consideration, it can be observed from single and multiple profile methods that the legacy system is a bottleneck, as far as execution times in the legacy are not drastically reduced. Therefore, we focus only on the legacy system and neglect all other local queues to construct an approximate queueing model (see Fig. 3). It is also observed that the system behaves more like multiple server queue rather than processor sharing for the parameters we examined. Summarising these arguments, we adopted a simple M/G/c queue as an approximation model and apply the following formula

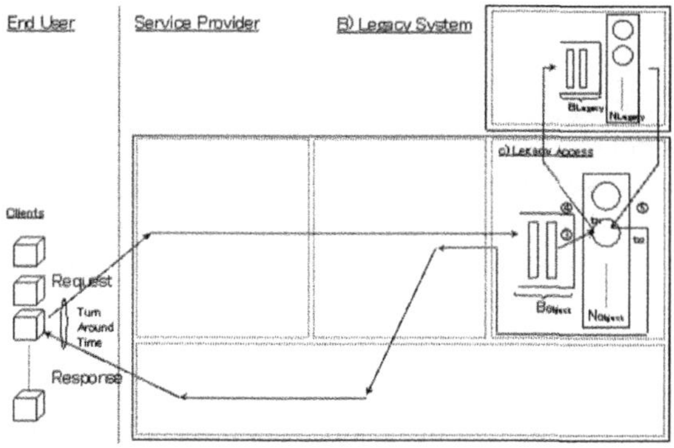

Fig. 3. An M/G/c queueing model for approximate analysis

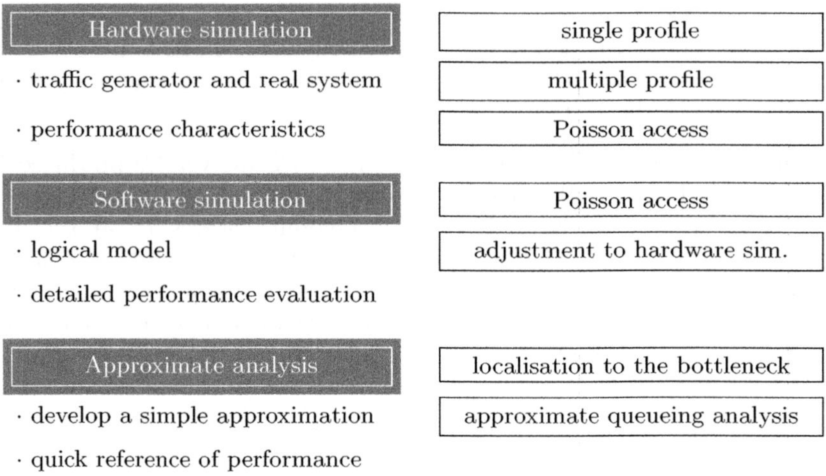

Fig. 4. A Diagram of three approaches for performance evaluation

for the mean waiting time given by Kimura [3] which is shown to provide reasonable approximations for a wide range of service time distributions.

$$W_q(\mathrm{M/G/c}) \approx \frac{1 + c_s^2}{\frac{2c_s^2}{W_q(\mathrm{M/M/c})} + \frac{2c_s^2}{W_q(\mathrm{M/D/c})}} \tag{1}$$

Here c_s is a coefficient of variation of the service time and $W_q(\cdot/\cdot/\mathrm{c})$ denotes the mean waiting time of $\cdot/\cdot/\mathrm{c}$ queue. It should be noted that an explicit formula is available for $W_q(\mathrm{M/M/c})$ and $W_q(\mathrm{M/D/c})$ can be obtained by a simple numerical algorithm. Therefore, if an arrival rate, mean and variance of execution times at the legacy system are at hand, then an approximate mean turn around time can be readily calculated from (1).

4 Experimental results

In this section, we will show some experimental results of our approach applied to evaluate the performance of the prototype system. In the first experiment, we assume constant execution time on the legacy system as in the case shown in Sect. 3. To evaluate the performance in a more realistic situation, random execution time is assumed in the second experiment whose distribution is estimated from the real data.

In both experiments, requests are supposed to arrive at the system according to a Poisson process. And the number of threads connecting with the legacy system is set to 5. Under such circumstances, we compare three approaches described in Sect. 3 to see the accuracy and properties of the evaluation.

Table 3 shows the expected turn around time for the constant execution time 2000 msec. on the legacy system. Arrival rate of the Poisson process is set to 1.0, 1.5 and 2.0 each of which corresponds to the traffic intensity 0.4, 0.6 and 0.8, respectively. A number given in parentheses shows a relative error compared with the hardware simulation. Also, 95% confidence intervals are shown for the hardware and software simulation. Since successive sojourn times are strongly correlated especially for high traffic intensity, we used so-called block method to calculate the confidence interval [8]. As can be seen in the table, software simulation overestimates the mean turn around time while the approximate analysis underestimates it.

Table 3. Mean turn around time (msec.) of the hardware simulation, software simulation and approximate analysis for constant execution time (2000 msec.) at the legacy system. A number given in parentheses shows a relative error (%) compared with the mean value of the hardware simulation. $\pm$ represents 95% confidence interval calculated by a block method

arrival rate (req./sec.)	hardware	software	approximation
1.0	2040 ± 3	2045 ± 4 (0.2%)	2025 (−0.7%)
1.5	2150 ± 9	2168 ± 13 (0.8%)	2132 (−0.8%)
2.0	2624 ± 44	2751 ± 60 (4.8%)	2578 (−1.8%)

To evaluate the system under more realistic circumstances, we next collect the real data of the execution time on the legacy system. We observed from the data that they are rather variable and have a heavier tail than exponential distribution which is commonly used within the field of performance evaluation. Among other specific distributions, the best fitted distribution based on Kolmogorov-Smirnov statistics was that constant plus log-normal distribution. The parameters of the distribution are given in (2) where X denotes a standard normal random variable. The mean and standard deviation of execution times are thus 4808 (msec.) and 1228 (msec.), respectively.

$$\text{(execution time)} = 875 + 1000e^{1.323+0.305X} \text{ (msec.)} \tag{2}$$

Table 4 shows the expected turn around time for three approaches. We set arrival rate as 0.5, 0.7 and 1.0 with corresponding traffic intensities 0.48, 0.67 and 0.96, respectively. In this case, both software simulation and approximate analysis overestimate the hardware simulation.

Summarising the above and some other experimental results, we observed the following.

- Overall, the software simulation and approximate analysis provide reasonable evaluation of the performance from practical viewpoint.
- The software simulation is apt to overestimate the hardware simulation. A plausible reason of this tendency is that the logical model for the software

Table 4. Mean turn around time (msec.) of the hardware simulation, software simulation and approximate analysis for random execution times at the legacy system whose distribution is given in (2) A number given in parentheses shows a relative error compared with the hardware simulation. ± represents 95% confidence interval calculated by a block method

arrival rate (req./sec.)	hardware	software	approximation
0.5	4923 ± 25	5015 ± 50 (1.9%)	4880 (−0.9%)
0.7	5342 ± 61	5582 ± 109 (4.5%)	5300 (−0.8%)
1.0	10071 ± 435	11590 ± 414 (15.1%)	11300 (12.2%)

simulation neglects some mechanisms equipped with the real system to reduce execution times. Examples of such mechanisms are time limit for the sojourn time in the system (a request is forcibly deleted from the system if it stays in the system for certain period of time), keep-alive mechanism (execution time at the connection is reduced if a request arrives when the former request is executed), and caching objects (reuse of objects to avoid object-creating overhead).

- The deviation from the hardware simulation increases as the traffic intensity increases. This is natural tendency commonly observed in the queueing approximation since the mean sojourn time rapidly increases as the traffic intensity exceeds certain threshold, say 0.7.

5 Summary

In this paper, we introduced a hybrid approach to evaluate the performance of a Web-client system and described our experience of its application to a prototype system. Since three types of approaches, hardware simulation, software simulation and approximate analysis are combined, it is possible for users to choose one of them according to necessary accuracy and their skills. For example, these approaches will be used in the following way in our case. A team developing the system first consult to research group to collect data of the system by the hardware simulation. Based on the data, the team implement the software simulation and determine the capacity and configuration of the system. Approximate analysis will be used as a quick reference when practitioners install and test the system at their clients.

To ensure the usefulness of the approach, we need more experiments to apply it to larger and more complex systems. Especially, a kind of experimental design will be necessary for the hardware simulation to construct an adequate logical model and estimate its parameters precisely. Another important extension will be to introduce admission control. Since arrival rate of requests from the Internet fluctuates over time, a request should be rejected when a congestion level of the system exceed certain threshold. We hope the approach will helpful to find an appropriate acceptance/rejection policy of the system.

Because of rapid expansion of the Internet businesses, it is urged to establish a new framework for capacity planning and design of configuration of Web C/S systems. This framework should be easy to implement, cost- and time-effective, and able to ensure quality of service. Although our approach is yet to be completed, we hope it contains useful ideas to establish such framework, a grammar of capacity planning of Web C/S systems.

References

1. Chao, X., Miyazawa, M. and Pinedo, M. (1999) Queueing Networks. John Wiley & Sons.
2. Gelenbe, E. and Mitrani, I. (1980) Analysis and Synthesis of Computer Systems. Academic Press.
3. Kimura, T. (1994) Approximations for multi-server queues: system interpolations. Queueing Systems, 17 (3-4), 347–382.
4. Kino, I. (2002) Queueing Networks. Asakura Syoten (in Japanese).
5. Menascé, D.A. and Almeida, V.A.F. (1998) Capacity Planning for Web Performance. Prentice Hall PTR.
6. Menascé, D.A. and Almeida, V.A.F. (2002) Capacity Planning for Web Services. Prentice Hall PTR.
7. Pritsker, A.A.B., O'Reilly, J.J. and LaVal, D.K. (1997) Simulation with Visual SLAM and Awesim. Systems Publishing and John Wiley.
8. Ripley, B.D. (1987) Stochastic Simulation. Wiely.
9. Standard Performance Evaluation Corporation (2002) SPECjAppServer2002 Documentation. `http://www.spec.org/osg/jAppServer`.
10. Sun Microsystems, Inc. Enterprise Java Beans 1.1 and 2.0. Specifications. `http://java.sun.com/products/ejb`

Polynomial Time Perfect Sampler for Discretized Dirichlet Distribution

Tomomi Matsui and Shuji Kijima

Department of Mathematical Informatics,
Graduate School of Information Science and Technology,
The University of Tokyo, Bunkyo-ku, Tokyo 113-8656, Japan.
http://www.simplex.t.u-tokyo.ac.jp/~tomomi/
kijima@simplex.t.u-tokyo.ac.jp

Summary. In this paper, we propose a perfect (exact) sampling algorithm according to a discretized Dirichlet distribution. The Dirichlet distribution appears as prior and posterior distribution for the multinomial distribution in many statistical methods in bioinformatics. Our algorithm is a monotone coupling from the past algorithm, which is a Las Vegas type randomized algorithm. We propose a new Markov chain whose limit distribution is a discretized Dirichlet distribution. Our algorithm simulates transitions of the chain $O(n^3 \ln \Delta)$ times where n is the dimension (the number of parameters) and $1/\Delta$ is the grid size for discretization. Thus the obtained bound does not depend on the magnitudes of parameters. In each transition, we need to sample a random variable according to a discretized beta distribution (2-dimensional Dirichlet distribution). To show the polynomiality, we employ the path coupling method carefully and show that our chain is rapidly mixing.

Key words: coupling from the past (CFTP), Dirichlet distribution, Markov chain Monte Carlo, path coupling, sampling algorithm.

1 Introduction

Computer simulation is becoming popular in many fields. This trend implies the necessity to develop a grammar of technology development whose core is computer simulation with statistical thinking. For simulating a statistical model, we need to sample objects according to a desired probability distribution. Sampling algorithms based on Markov chain are useful techniques when a domain of the probability distribution has a combinatorial structure. This paper deals with a (discretized) Dirichlet distribution appearing in statistical models in bioinformatics. We propose a polynomial time perfect (exact)

* Contributed paper: received date: 29-Jun-05, accepted: 15-Nov-05

sampling algorithm according to a discretized Dirichlet distribution. Our algorithm is a monotone coupling from the past (CFTP) algorithm, which is an outcome of an interdisciplinary approach based on Markov chain theory and techniques of Las Vegas type randomized algorithms. The polynomiality of our perfect sampling algorithm supports to utilize simulation based algorithms in bioinformatics.

In this paper, we propose a new Markov chain for generating random samples according to a discretized Dirichlet distribution. In each transition of our Markov chain, we need a random sample according to a discretized beta distribution (2-dimensional Dirichlet distribution). Our algorithm simulates transitions of the Markov chain $\mathrm{O}(n^3 \ln \Delta)$ times in expectation where n is the dimension (the number of parameters) and $1/\Delta$ is the grid size for discretization. To show the polynomiality, we employ the path coupling method and show that the mixing rate of our chain is bounded by $n(n-1)^2(1+\ln(n(\Delta-n)/2))$.

Statistical methods are widely studied in bioinformatics since they are powerful tools to discover genes causing a (common) disease from a number of observed data. These methods often use EM algorithm, Markov chain Monte Carlo method, Gibbs sampler, and so on. The Dirichlet distribution appears as prior and posterior distribution for the multinomial distribution in these methods since the Dirichlet distribution is the conjugate prior of parameters of the multinomial distribution [17].

For example, Niu, Qin, Xu, and Liu proposed a Bayesian haplotype inference method [13], which decides phased individual genotypes (diplotype configuration for each subject) probabilistically. Another example is a population structure inferring algorithm by Pritchard, Stephens, and Donnely [14]. Their algorithm is based on MCMC method. In these examples, the Dirichlet distribution appears with various dimensions and various parameters. Thus we need an efficient algorithm for sampling from the Dirichlet distribution with arbitrary dimensions and parameters. The Dirichlet distribution was also used to estimate inbreeding coefficient and effective size from allele frequency changes [10], and to perform a meta-analysis of studies on the association of polymorphisms and risk of a disease [4]. In the paper [3], Burr used the distribution to examine the quasi-equilibrium theory for the distribution of rare alleles in a subdivided population. Kitada, Hayashi and Kishino used the Dirichlet distribution to estimate genetic distance between populations and effective population size [9]. To approximate the conditional genotypic probabilities for microsatellite loci for forensic medicine, Graham, Curran, and Weir used the distribution [7].

One approach of sampling from a (continuous) Dirichlet distribution is by rejection (see [6] for example). Even in the case that $n = 2$ (a gamma distribution), the ratio of rejection becomes higher as the parameter is smaller. Thus, it does not seem effective when the magnitudes of parameters are small. Another approach is a descretization of the domain and adoption of Metropolis-Hastings algorithm. Recently, Matsui, Motoki and Kamatani proposed a Markov chain for sampling from discretized Dirichlet distribution [11].

The mixing time of their chain is bounded by $(1/2)n(n-1)\ln((\Delta - n)/\varepsilon)$. The computational complexity of each transition of our chain is equivalent to that of the approximate sampler in [11]. Thus, when $\ln(1/\varepsilon)$ is greater than $O(n \ln \Delta)$, the computational time of our algorithm ($O(n^3 \ln \Delta)$) is competitive with or faster than that of the approximate sampler and the obtained random vector is exactly distributed according to the stationary distribution.

Propp and Wilson devised a surprising algorithm, called (monotone) CFTP algoritm (or backward coupling), which produces exact samples from the limit distribution [15, 16]. Monotone CFTP algorithm simulates infinite time transitions of a monotone Markov chain in a (probabilistically) finite time. To employ their result, we propose a new monotone Markov chain whose stationary distribution is a discretized Dirichlet distribution. One of the great advantages of perfect sampling algorithm is that the algorithm outputs a random vector exactly distributed according to the stationary distribution. Thus users never need to determine the error rate ε. If we need a sample according to highly accurate distribution, a perfect sampler becomes faster than any approximate sampler based on a Markov chain. Additionally, any over-estimation of the complexity does not affect the real computational time of a CFTP algorithm. On the other hand, if we employ an approximate sampler based on a Markov chain, the required number of iterations is exactly the same with the obtained mixing time, thereby any over-estimation of mixing time directly increase the practical computational time.

This paper is deeply related to authors' recent paper [8] which proposes a polynomial time perfect sampling algorithm for two-rowed contingency tables. The idea of the algorithm and outline of the proof described in this paper is similar to those in our paper [8]. The main difference of these two papers is that when we deal with the uniform distribution on two-rowed contingency tables, the transition probabilities of the Markov chain are predetermined. However, in case of discretized Dirichlet distribution, the transition probabilities vary with respect to the magnitude of paprameters. Thus, to show the monotonicity and polynomiality of the algorithm proposed in this paper, we need a detailed discussions which appear in Appendix section. Showing two lemmas in Appendix section is easier in case of the uniform distribution on two-rowed contingency tables.

2 Review of Coupling From the Past Algorithm

When we simulate an ergodic Markov chain for infinite time, we can gain a sample exactly according to the stationary distribution. Suppose that there exists a chain from infinite past, then a possible state at the present time of the chain for which we can have an evidence of the uniqueness without respect to an initial state of the chain, is a realization of a random sample exactly from the stationary distribution. This is the key idea of CFTP.

Suppose that we have an ergodic Markov chain MC with finite state space Ω and transition matrix P. The transition rule of the Markov chain $X \mapsto X'$ can be described by a deterministic function $\phi\colon \Omega \times [0,1) \to \Omega$, called *update function*, as follows. Given a random number Λ uniformly distributed over $[0,1)$, update function ϕ satisfies that $\Pr(\phi(x,\Lambda) = y) = P(x,y)$ for any x, $y \in \Omega$. We can realize the Markov chain by setting $X' = \phi(X,\Lambda)$. Clearly, update function corresponding to the given transition matrix P is not unique. The result of transitions of the chain from the time t_1 to t_2 ($t_1 < t_2$) with a sequence of random numbers $\boldsymbol{\lambda} = (\lambda[t_1], \lambda[t_1+1], \ldots, \lambda[t_2-1]) \in [0,1)^{t_2-t_1}$ is denoted by $\Phi_{t_1}^{t_2}(x,\boldsymbol{\lambda})\colon \Omega \times [0,1)^{t_2-t_1} \to \Omega$ where $\Phi_{t_1}^{t_2}(x,\boldsymbol{\lambda}) \stackrel{\text{def.}}{=} \phi(\phi(\cdots(\phi(x,\lambda[t_1]),\ldots,\lambda[t_2-2]),\lambda[t_2-1])$. We say that a sequence $\lambda \in [0,1)^{|T|}$ satisfies the *coalescence condition*, when $\exists y \in \Omega$, $\forall x \in \Omega$, $y = \Phi_T^0(x,\boldsymbol{\lambda})$.

With these preparation, standard Coupling From The Past algorithm is expressed as follows.

CFTP Algorithm [15]

Step 1. Set the starting time period $T := -1$ to go back, and set $\boldsymbol{\lambda}$ be the empty sequence.

Step 2. Generate random real numbers $\lambda[T]$, $\lambda[T+1]$, …, $\lambda[\lceil T/2 \rceil - 1] \in [0,1)$, and insert them to the head of $\boldsymbol{\lambda}$ in order, i.e., put $\boldsymbol{\lambda} := (\lambda[T], \lambda[T+1], \ldots, \lambda[-1])$.

Step 3. Start a chain from each element $x \in \Omega$ at time period T, and run each chain to time period 0 according to the update function ϕ with the sequence of numbers in $\boldsymbol{\lambda}$. (Note that every chain uses the common sequence $\boldsymbol{\lambda}$.)

Step 4. [Coalescence check]
- (a) If $\exists y \in \Omega$, $\forall x \in \Omega$, $y = \Phi_T^0(x,\boldsymbol{\lambda})$, then return y and stop.
- (b) Else, update the starting time period $T := 2T$, and go to Step 2.

Theorem 1 (CFTP Theorem [15]). *Let MC be an ergodic finite Markov chain with state space Ω, defined by an update function $\phi\colon \Omega \times [0,1) \to \Omega$. If the CFTP algorithm terminates with probability 1, then the obtained value is a realization of a random variable exactly distributed according to the stationary distribution.*

Theorem 1 gives a (probabilistically) finite time algorithm for infinite time simulation. However, simulations from all states executed in Step 3 are hard requirements.

Suppose that there exists a partial order "$\succeq$" on the set of states Ω. A transition rule expressed by a deterministic update function ϕ is called *monotone* (with respect to "$\succeq$") if $\forall \lambda \in [0,1)$, $\forall x$, $\forall y \in \Omega$, $x \succeq y \Rightarrow \phi(x,\lambda) \succeq \phi(y,\lambda)$. For ease, we also say that a chain is *monotone* if the chain has a *monotone* transition rule.

Theorem 2 (monotone CFTP [15, 5]). *Suppose that a Markov chain defined by an update function ϕ is monotone with respect to a partially ordered* *set of states* $(\Omega, \succeq)$, *and* $\exists x_{\max}$, $\exists x_{\min} \in \Omega$, $\forall x \in \Omega$, $x_{\max} \succeq x \succeq x_{\min}$. *Then the CFTP algorithm terminates with probability 1, and a sequence* $\lambda \in [0,1)^{|T|}$ *satisfies the coalescence condition, i.e.,* $\exists y \in \Omega$, $\forall x \in \Omega$, $y = \Phi_T^0(x, \boldsymbol{\lambda})$, *if and only if* $\Phi_T^0(x_{\max}, \boldsymbol{\lambda}) = \Phi_T^0(x_{\min}, \boldsymbol{\lambda})$.

When the given Markov chain satisfies the conditions of Theorem 2, we can modify CFTP algorithm by substituting Step 4 (a) by

Step 4. (a)′ If $\exists y \in \Omega$, $y = \Phi_T^0(x_{\max}, \boldsymbol{\lambda}) = \Phi_T^0(x_{\min}, \boldsymbol{\lambda})$, then return y and stop.

The algorithm obtained by the above modification is called a monotone CFTP algorithm.

3 Perfect Sampling Algorithm

In this paper, we denote the set of integers (non-negative or positive integers) by Z (Z_+, Z_{++}). Dirichlet random vector $P = (P_1, P_2, \ldots, P_n)$ with non-negative parameters $u_1, \ldots, u_n$ is a vector of random variables that admits the probability density function

$$\frac{\Gamma(\sum_{i=1}^n u_i)}{\prod_{i=1}^n \Gamma(u_i)} \prod_{i=1}^n p_i^{u_i - 1}$$

defined on the set $\{(p_1, p_2, \ldots, p_n) \in \mathrm{R}^n \mid p_1 + \cdots + p_n = 1,\ p_1, p_2, \ldots, p_n > 0\}$ where $\Gamma(u)$ is the gamma function. Throughout this paper, we assume that $n \geq 2$.

For any integer $\Delta \geq n$, we discretize the domain with grid size $1/\Delta$ and obtain a discrete set of integer vectors Ω defined by

$$\Omega \stackrel{\text{def.}}{=} \{(x_1, x_2, \ldots, x_n) \in \mathrm{Z}_{++}^n \mid x_i > 0\ (\forall i),\ x_1 + \cdots + x_n = \Delta\}.$$

A discretized Dirichlet random vector with non-negative parameters $u_1, \ldots, u_n$ is a random vector $X = (X_1, \ldots, X_n) \in \Omega$ with the distribution

$$\Pr[X = (x_1, \ldots, x_n)] \stackrel{\text{def.}}{=} C_\Delta \prod_{i=1}^n (x_i/\Delta)^{u_i - 1}$$

where C_Δ is the partition function (normarizing constant) defined by $(C_\Delta)^{-1} \stackrel{\text{def.}}{=} \sum_{\boldsymbol{x} \in \Omega} \prod_{i=1}^n (x_i/\Delta)^{u_i - 1}$.

For any integer $b \geq 2$, we introduce a set of 2-dimensional integer vectors $\Omega(b) \stackrel{\text{def.}}{=} \{(Y_1, Y_2) \in \mathrm{Z}^2 \mid Y_1,\ Y_2 > 0,\ Y_1 + Y_2 = b\}$ and a distribution function $f_b(Y_1, Y_2 \mid u_i, u_j)$: $\Omega(b) \to [0,1]$ with non-negative parameters u_i, u_j defined by

$$f_b(Y_1, Y_2 \mid u_i, u_j) \stackrel{\text{def.}}{=} C(u_i, u_j, b) Y_1^{u_i - 1} Y_2^{u_j - 1}$$

where $(C(u_i, u_j, b))^{-1} \stackrel{\text{def.}}{=} \sum_{(Y_1,Y_2)\in\Omega(b)} Y_1^{u_i-1} Y_2^{u_j-1}$ is the partition function. We also introduce a vector $(g_b(0 \mid u_i, u_j), g_b(1 \mid u_i, u_j), \ldots, g_b(b-1 \mid u_i, u_j))$ defined by

$$g_b(k \mid u_i, u_j) \stackrel{\text{def.}}{=} \begin{cases} 0 & (k = 0) \\ \sum_{l=1}^{k} C(u_i, u_j, b) l^{u_i-1} (b-l)^{u_j-1} & (k \in \{1, 2, \ldots, b-1\}) \end{cases}$$

It is clear that $0 = g_b(0 \mid u_i, u_j) < g_b(1 \mid u_i, u_j) < \cdots < g_b(b-1 \mid u_i, u_j) = 1$.

We describe our Markov chain $\mathcal{M}$ with state space Ω. Given a current state $X \in \Omega$, we generate a random number $\lambda \in [1, n)$. Then transition $X \mapsto X'$ with respect to λ takes place as follows.

Markov chain $\mathcal{M}$
Input: A random number $\lambda \in [1, n)$.
Step 1: Put $i := \lfloor \lambda \rfloor$ and $b := X_i + X_{i+1}$.
Step 2: Let $k \in \{1, 2, \ldots, b-1\}$ be a unique value satisfying

$$g_b(k-1 \mid u_i, u_{i+1}) \le (\lambda - \lfloor \lambda \rfloor) < g_b(k \mid u_i, u_{i+1}).$$

Step 3: Put $X'_j := \begin{cases} k & (j = i), \\ b-k & (j = i+1), \\ X_j & (\text{otherwise}). \end{cases}$

The update function $\phi\colon \Omega \times [1, n) \to \Omega$ of our chain is defined by $\phi(X, \lambda) \stackrel{\text{def.}}{=} X'$ where X' is determined by the above procedure. Clearly, this chain is irreducible and aperiodic. Since the detailed balance equations hold, the stationary distribution of the above Markov chain $\mathcal{M}$ is the discretized Dirichlet distribution.

We define two special states $X_{\mathrm{U}}, X_{\mathrm{L}} \in \Omega$ by

$$X_{\mathrm{U}} \stackrel{\text{def.}}{=} (\Delta - n + 1, 1, 1, \ldots, 1), \quad X_{\mathrm{L}} \stackrel{\text{def.}}{=} (1, 1, \ldots, 1, \Delta - n + 1).$$

Now we describe our algorithm.

Algorithm 1
Step 1. Set the starting time period $T := -1$ to go back, and $\boldsymbol{\lambda}$ be the empty sequence.
Step 2. Generate random real numbers $\lambda[T], \lambda[T+1], \ldots, \lambda[\lceil T/2 \rceil - 1] \in [1, n)$ and put $\boldsymbol{\lambda} = (\lambda[T], \lambda[T-1], \ldots, \lambda[-1])$.
Step 3. Start two chains from X_{U} and X_{L}, respectively at time period T, and run them to time period 0 according to the update function ϕ with the sequence of numbers in $\boldsymbol{\lambda}$.
Step 4. [Coalescence check]
If $\exists Y \in \Omega$, $Y = \Phi_T^0(X_{\mathrm{U}}, \boldsymbol{\lambda}) = \Phi_T^0(X_{\mathrm{L}}, \boldsymbol{\lambda})$, then return Y and stop.
Else, update the starting time period $T := 2T$, and go to Step 2.

Theorem 3. *With probability 1, Algorithm 1 terminates and returns a state. The state obtained by Algorithm 1 is a realization of a sample exactly according to the discretized Dirichlet distribution.*

The above theorem guarantees that Algorithm 1 is a perfect sampling algorithm. We prove the above theorem by showing the monotonicity in the next section.

4 Monotonicity of the Chain

In Section 2, we described two theorems. Thus we only need to show the monotonicity of our chain. First, we introduce a partial order on the state space Ω.

For any vector $X \in \Omega$, we define the *cumulative sum vector* $\mathrm{c}_X \in \mathrm{Z}_+^{n+1}$ by

$$\mathrm{c}_X(i) \stackrel{\text{def.}}{=} \begin{cases} 0 & (i = 0), \\ X_1 + X_2 + \cdots + X_i & (i \in \{1, 2, \ldots, n\}), \end{cases}$$

where $\mathrm{c}_X = (\mathrm{c}_X(0), \mathrm{c}_X(1), \ldots, \mathrm{c}_X(n))$. Clearly, there exists a bijection between Ω and the set $\{\mathrm{c}_X \mid X \in \Omega\}$. For any pair of states $X, Y \in \Omega$, we say $X \succeq Y$ if and only if $\mathrm{c}_X \geq \mathrm{c}_Y$. It is clear that the relation "$\succeq$" is a partial order on Ω. We can see easily that $\forall X \in \Omega$, $X_{\mathrm{U}} \succeq X \succeq X_{\mathrm{L}}$.

We say that a state $X \in \Omega$ *covers* $Y \in \Omega$ (at j), denoted by $X \succ\!\!\cdot\, Y$ (or $X \succ\!\!\cdot_j\, Y$), when

$$\mathrm{c}_X(i) - \mathrm{c}_Y(i) = \begin{cases} 1 & (i = j), \\ 0 & (\text{otherwise}). \end{cases}$$

Note that $X \succ\!\!\cdot_j\, Y$ if and only if

$$X_i - Y_i = \begin{cases} +1 & (i = j), \\ -1 & (i = j + 1), \\ 0 & (\text{otherwise}). \end{cases}$$

Next, we show a key lemma for proving monotonicity.

Lemma 1. $\forall X$, $\forall Y \in \Omega$, $\forall \lambda \in [1, n)$, $X \succ\!\!\cdot_j\, Y \Rightarrow \phi(X, \lambda) \succeq \phi(Y, \lambda)$.

Proof. We denote $\phi(X, \lambda)$ by X' and $\phi(Y, \lambda)$ by Y' for simplicity. For any index $i \neq \lfloor \lambda \rfloor$, it is easy to see that $\mathrm{c}_X(i) = \mathrm{c}_{X'}(i)$ and $\mathrm{c}_Y(i) = \mathrm{c}_{Y'}(i)$, and so $\mathrm{c}_{X'}(i) - \mathrm{c}_{Y'}(i) = \mathrm{c}_X(i) - \mathrm{c}_Y(i) \geq 0$ since $X \succeq Y$. In the following, we show that $\mathrm{c}_{X'}(\lfloor \lambda \rfloor) \geq \mathrm{c}_{Y'}(\lfloor \lambda \rfloor)$.

Clearly from the definition of our Markov chain, $X'_{\lfloor \lambda \rfloor}$ is a unique value k' satisfying

$$g_{b'}(k' - 1 \mid u_{\lfloor \lambda \rfloor}, u_{\lfloor \lambda \rfloor + 1}) \leq (\lambda - \lfloor \lambda \rfloor) < g_{b'}(k' \mid u_{\lfloor \lambda \rfloor}, u_{\lfloor \lambda \rfloor + 1})$$

where $b' \stackrel{\text{def.}}{=} X_{\lfloor \lambda \rfloor} + X_{\lfloor \lambda \rfloor + 1}$. Similarly, $Y'_{\lfloor \lambda \rfloor}$ is a unique value k'' satisfying

$$g_{b''}(k''-1\,|\,u_{\lfloor\lambda\rfloor},u_{\lfloor\lambda\rfloor+1}) \leq (\lambda-\lfloor\lambda\rfloor) < g_{b''}(k''\,|\,u_{\lfloor\lambda\rfloor},u_{\lfloor\lambda\rfloor+1})$$

where $b'' \stackrel{\text{def.}}{=} Y_{\lfloor\lambda\rfloor}+Y_{\lfloor\lambda\rfloor+1}$. We need to consider following three cases.

Case 1: If $\lfloor\lambda\rfloor \neq j-1$ and $\lfloor\lambda\rfloor \neq j+1$, then $b'=b''$ and so we have $X'_{\lfloor\lambda\rfloor}=k'=k''=Y'_{\lfloor\lambda\rfloor}$.

Case 2: Consider the case that $\lfloor\lambda\rfloor = j-1$. Since $X \succ_j Y$, we have $b'=b''+1$. From the definition of cumulative sum vector,

$$\begin{aligned}&\mathrm{c}_{X'}(j-1)-\mathrm{c}_{Y'}(j-1)=\mathrm{c}_{X'}(j-2)+X'_{j-1}-\mathrm{c}_{Y'}(j-2)-Y'_{j-1}\\&\quad=\mathrm{c}_{X}(j-2)+X'_{j-1}-\mathrm{c}_{Y}(j-2)-Y'_{j-1}=X'_{j-1}-Y'_{j-1}.\end{aligned}$$

Thus, it is enough to show that $X'_{j-1}\geq Y'_{j-1}$.

Lemma 5 in Appendix section implies the following inequalities,

$$\begin{aligned}0&=g_{b''+1}(0\,|\,u_{j-1},u_j)=g_{b''}(0\,|\,u_{j-1},u_j)\leq g_{b''+1}(1\,|\,u_{j-1},u_j)\\&\leq g_{b''}(1\,|\,u_{j-1},u_j)\leq\cdots\leq g_{b''+1}(k-1\,|\,u_{j-1},u_j)\leq g_{b''}(k-1\,|\,u_{j-1},u_j)\\&\leq g_{b''+1}(k\,|\,u_{j-1},u_j)\leq\cdots\leq g_{b''+1}(b''-1\,|\,u_{j-1},u_j)\leq g_{b''}(b''-1\,|\,u_{j-1},u_j)\\&=g_{b''+1}(b''\,|\,u_{j-1},u_j)=1,\end{aligned}$$

which we will call *alternating inequalities*. For example, if inequalities

$$g_{b''+1}(k-1\,|\,u_{j-1},u_j)\leq(\lambda-\lfloor\lambda\rfloor)<g_{b''}(k-1\,|\,u_{j-1},u_j)\leq g_{b''+1}(k\,|\,u_{j-1},u_j)$$

hold, then $X'_{\lfloor\lambda\rfloor}=k>k-1=Y'_{\lfloor\lambda\rfloor}$. And if

$$g_{b''+1}(k-1\,|\,u_{j-1},u_j)\leq g_{b''}(k-1\,|\,u_{j-1},u_j)\leq(\lambda-\lfloor\lambda\rfloor)<g_{b''+1}(k\,|\,u_{j-1},u_j)$$

hold, then $X'_{\lfloor\lambda\rfloor}=k=Y'_{\lfloor\lambda\rfloor}$. Thus we have

$$\begin{pmatrix}X'_{j-1}\\Y'_{j-1}\end{pmatrix}\in\left\{\begin{pmatrix}1\\1\end{pmatrix},\begin{pmatrix}2\\1\end{pmatrix},\begin{pmatrix}2\\2\end{pmatrix},\begin{pmatrix}3\\2\end{pmatrix},\ldots,\begin{pmatrix}b''-1\\b''-1\end{pmatrix},\begin{pmatrix}b''\\b''-1\end{pmatrix}\right\}$$

(see Figure 1). From the above we have that $X'_{j-1}\geq Y'_{j-1}$.

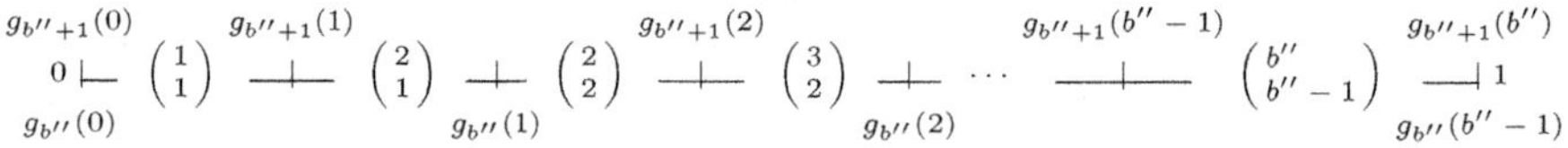

Fig. 1. A figure of alternating inequalities. In the above, we denote $g_{b''}(k\,|\,u_{j-1},u_j)$ and $g_{b''+1}(k\,|\,u_{j-1},u_j)$ by $g_{b''}(k)$, $g_{b''+1}(k)$, respectively.

Case 3: Consider the case that $\lfloor\lambda\rfloor = j+1$. Since $X \succ_j Y$, we have $b'+1=b''$. From the definition of cumulative sum vector,

$$\begin{aligned}&\mathrm{c}_{X'}(j+1)-\mathrm{c}_{Y'}(j+1)=\mathrm{c}_{X'}(j)+X'_{j+1}-\mathrm{c}_{Y'}(j)-Y'_{j+1}\\&\quad=\mathrm{c}_{X}(j)+X'_{j+1}-\mathrm{c}_{Y}(j)-Y'_{j+1}=1+X'_{j+1}-Y'_{j+1}.\end{aligned}$$

Thus, it is enough to show that $1 + X'_{j+1} \geq Y'_{j+1}$.

Then Lemma 5 also implies the following alternating inequalities

$$\begin{aligned}0 &= g_{b'+1}(0 \,|\, u_{j+1}, u_{j+2}) = g_{b'}(0 \,|\, u_{j+1}, u_{j+2})\\ &\leq g_{b'+1}(1 \,|\, u_{j+1}, u_{j+2}) \leq g_{b'}(1 \,|\, u_{j+1}, u_{j+2}) \leq \cdots\\ &\leq g_{b'+1}(k-1 \,|\, u_{j+1}, u_{j+2}) \leq g_{b'}(k-1 \,|\, u_{j+1}, u_{j+2}) \leq g_{b'+1}(k \,|\, u_{j+1}, u_{j+2}) \leq \cdots\\ &\leq g_{b'+1}(b'-1 \,|\, u_{j+1}, u_{j+2}) \leq g_{b'}(b'-1 \,|\, u_{j+1}, u_{j+2}) = g_{b'+1}(b' \,|\, u_{j+1}, u_{j+2}) = 1.\end{aligned}$$

Then it is easy to see that

$$\binom{X'_{j+1}}{Y'_{j+1}} \in \left\{ \binom{1}{1}, \binom{1}{2}, \binom{2}{2}, \binom{2}{3}, \ldots, \binom{b'-1}{b'-1}, \binom{b'-1}{b'} \right\}$$

(see Figure 2). From the above, we obtain the inequality $1 + X'_{j+1} \geq Y'_{j+1}$. □

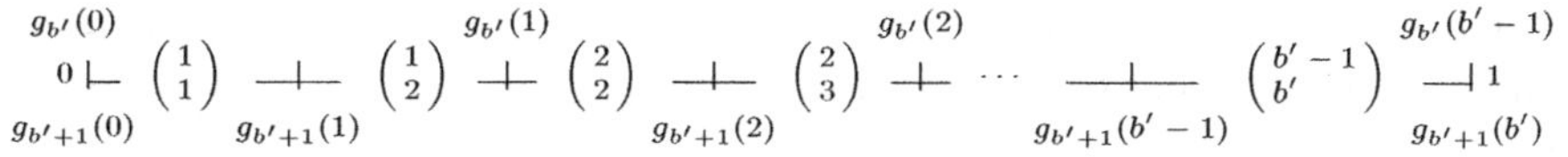

Fig. 2. A figure of alternating inequalities. In the above, we denote $g_{b'}(k \,|\, u_{j+1}, u_{j+2})$ and $g_{b'+1}(k \,|\, u_{j+1}, u_{j+2})$ by $g_{b'}(k)$, $g_{b'+1}(k)$, respectively.

Lemma 2. *The Markov chain $\mathcal{M}$ defined by the update function ϕ is monotone with respect to "$\succeq$", i.e., $\forall \lambda \in [1, n)$, $\forall X$, $\forall Y \in \Omega$, $X \succeq Y \Rightarrow \phi(X, \lambda) \succeq \phi(Y, \lambda)$.*

Proof. It is easy to see that there exists a sequence of states $Z_1, Z_2, \ldots, Z_r$ with appropreate length satisfying $X = Z_1 \succ\!\!\!\cdot\; Z_2 \succ\!\!\!\cdot\; \cdots \succ\!\!\!\cdot\; Z_r = Y$. Then applying Lemma 1 repeatedly, we can show that $\phi(X, \lambda) = \phi(Z_1, \lambda) \succeq \phi(Z_2, \lambda) \succeq \cdots \succeq \phi(Z_r, \lambda) = \phi(Y, \lambda)$. □

Lastly, we show the correctness of our algorithm.

Proof of Theorem 3. From Lemma 2, the Markov chain is monotone, and it is clear that $(X_{\mathrm{U}}, X_{\mathrm{L}})$ is a unique pair of maximum and minimum states. Thus Algorithm 1 is a monotone CFTP algorithm and Theorems 1 and 2 implies Theorem 3. □

5 Expected Running Time

Here, we discuss the running time of our algorithm. In this section, we assume the following.

Condition 1 *Parameters are arranged in non-increasing order, i.e.,* $u_1 \geq u_2 \geq \cdots \geq u_n$.

The following is a main result of this paper.

Theorem 4. *Under Condition 1, the expected number of tansitions executed in Algorithm 1 is bounded by* $\mathrm{O}(n^3 \ln \Delta)$ *where* n *is the dimension (number of parameters) and* $1/\Delta$ *is the grid size for discretization.*

In the rest of this section, we prove Theorem 4 by estimating the expectation of *coalescence time* $T_* \in \mathrm{Z}_{++}$ defined by $T_* \stackrel{\text{def.}}{=} \min\{t > 0 \mid \exists y \in \Omega, \forall x \in \Omega,\ y = \Phi^0_{-t}(x, \boldsymbol{\Lambda})\}$. Note that T_* is a random variable.

Given a pair of probabilistic distributions ν_1 and ν_2 on the finite state space Ω, the *total variation distance* between ν_1 and ν_2 is defined by $\mathrm{D_{TV}}(\nu_1, \nu_2) \stackrel{\text{def.}}{=} \frac{1}{2}\sum_{x\in\Omega} |\nu_1(x) - \nu_2(x)|$. The *mixing rate* of an ergodic Markov chain is defined by $\tau \stackrel{\text{def.}}{=} \max_{x\in\Omega}\{\min\{t \mid \forall s \geq t,\ \mathrm{D_{TV}}(\pi, P^s_x) \leq 1/\mathrm{e}\}\}$ where π is the stationary distribution and P^s_x is the probabilistic distribution of the chain at time s with initial state x. Path Coupling Theorem is a useful technique for bounding the mixing rate.

Theorem 5 (Path Coupling [1, 2]). *Let* MC *be a finite ergodic Markov chain with state space* Ω*. Let* $G = (\Omega, \mathcal{E})$ *be a connected undirected graph with vertex set* Ω *and edge set* $\mathcal{E} \subseteq \binom{\Omega}{2}$*. Let* $l \colon \mathcal{E} \to \mathrm{R}$ *be a positive length defined on the edge set. For any pair of vertices* $\{x, y\}$ *of* G*, the distance between* x *and* y*, denoted by* $d(x, y)$ *and/or* $d(y, x)$*, is the length of a shortest path between* x *and* y*, where the length of a path is the sum of the lengths of edges in the path. Suppose that there exists a joint process* $(X, Y) \mapsto (X', Y')$ *with respect to* MC *satisfying that whose marginals are a faithful copy of* MC *and*

$$0 < \exists \beta < 1,\ \forall \{X, Y\} \in \mathcal{E}, \quad \mathrm{E}[d(X', Y')] \leq \beta d(X, Y).$$

Then the mixing rate τ *of Markov chain* MC *satisfies* $\tau \leq (1 - \beta)^{-1} (1 + \ln(D/d))$*, where* $d \stackrel{\text{def.}}{=} \min_{(x,y)\in\Omega^2} d(x, y)$ *and* $D \stackrel{\text{def.}}{=} \max_{(x,y)\in\Omega^2} d(x, y)$.

The above theorem differs from the original theorem in [1] since the integrality of the edge length is not assumed. We drop the integrality and introduced the minimum distance d. This modification is not essential and we can show Theorem 5 similarly.

Now, we show the polynomiality of our algorithm. First, we estimate the mixing rate of our chain $\mathcal{M}$ by employing Path Coupling Theorem. In the proof of the following lemma, Condition 1 plays an important role.

Lemma 3. *Under Condition 1, the mixing rate* τ *of our Markov chain* $\mathcal{M}$ *satisfies* $\tau \leq n(n-1)^2(1 + \ln(n(\Delta - n)/2))$.

Proof. Let $G = (\Omega, \mathcal{E})$ be an undirected simple graph with vertex set Ω and edge set $\mathcal{E}$ defined as follows. A pair of vertices $\{X, Y\}$ is an edge if and only if $(1/2)\sum_{i=1}^{n} |X_i - Y_i| = 1$. Clearly, the graph G is connected. For each edge $e = \{X, Y\} \in \mathcal{E}$, there exists a unique pair of indices $j_1, j_2 \in \{1, \ldots, n\}$, called the *supporting pair* of e, satisfying

$$|X_i - Y_i| = \begin{cases} 1 & (i = j_1, j_2), \\ 0 & (\text{otherwise}). \end{cases}$$

We define the length $l(e)$ of an edge e by $l(e) \overset{\text{def.}}{=} (1/(n-1))\sum_{i=1}^{j^*-1}(n-i)$ where $j^* = \max\{j_1, j_2\} \geq 2$ and $\{j_1, j_2\}$ is the supporting pair of e. Note that $1 \leq \min_{e \in \mathcal{E}} l(e) \leq \max_{e \in \mathcal{E}} l(e) \leq n/2$. For each pair $X, Y \in \Omega$, we define the distance $d(X, Y)$ be the length of the shortest path between X and Y on G. Clearly, the diameter of G, i.e., $\max_{(X,Y) \in \Omega^2} d(X, Y)$, is bounded by $n(\Delta - n)/2$, since $d(X, Y) \leq (n/2)\sum_{i=1}^{n}(1/2)|X_i - Y_i| \leq (n/2)(\Delta - n)$ for any $(X, Y) \in \Omega^2$. The definition of edge length implies that for any edge $\{X, Y\} \in \mathcal{E}$, $d(X, Y) = l(\{X, Y\})$.

We define a joint process $(X, Y) \mapsto (X', Y')$ by $(X, Y) \mapsto (\phi(X, \Lambda), \phi(Y, \Lambda))$ with uniform real random number $\Lambda \in [1, n)$ where ϕ is the update function defined in Section 3. Now we show that

$$\mathrm{E}[d(X', Y')] \leq \beta d(X, Y), \quad \beta = 1 - 1/(n(n-1)^2), \tag{1}$$

for any pair $\{X, Y\} \in \mathcal{E}$. In the following, we denote the supporting pair of $\{X, Y\}$ by $\{j_1, j_2\}$. Without loss of generality, we can assume that $j_1 < j_2$, and $X_{j_2} + 1 = Y_{j_2}$.

Case 1: When $\lfloor \Lambda \rfloor = j_2 - 1$, we will show that

$$\mathrm{E}[d(X', Y') \,|\, \lfloor \Lambda \rfloor = j_2 - 1] \leq d(X, Y) - (1/2)(n - j_2 + 1)/(n - 1).$$

In case $j_1 = j_2 - 1$, $X' = Y'$ with conditional probabilty 1. Hence $d(X', Y') = 0$. In the following, we consider the case $j_1 < j_2 - 1$. Put $b' = X_{j_2-1} + X_{j_2}$ and $b'' = Y_{j_2-1} + Y_{j_2}$. Since $X_{j_2} + 1 = Y_{j_2}$, $b' + 1 = b''$ holds. From the definition of the update function of our Marokov chain, we have followings,

$$\begin{aligned} X'_{j_2-1} &= k \Leftrightarrow [g_{b'}(k-1 \,|\, u_{j_2-1}, u_{j_2}) \leq \Lambda - \lfloor \Lambda \rfloor < g_{b'}(k \,|\, u_{j_2-1}, u_{j_2})] \\ Y'_{j_2-1} &= k \Leftrightarrow [g_{b'+1}(k-1 \,|\, u_{j_2-1}, u_{j_2}) \leq \Lambda - \lfloor \Lambda \rfloor < g_{b'+1}(k \,|\, u_{j_2-1}, u_{j_2})]. \end{aligned}$$

As described in the proof of Lemma 1, the alternating inequalities

$$\begin{aligned} 0 = g_{b'+1}(0 \,|\, u_{j_2-1}, u_{j_2}) &= g_{b'}(0 \,|\, u_{j_2-1}, u_{j_2}) \\ \leq g_{b'+1}(1 \,|\, u_{j_2-1}, u_{j_2}) &\leq g_{b'}(1 \,|\, u_{j_2-1}, u_{j_2}) \leq \cdots \\ \leq g_{b'+1}(b'-1 \,|\, u_{j_2-1}, u_{j_2}) &\leq g_{b'}(b'-1 \,|\, u_{j_2-1}, u_{j_2}) = g_{b'+1}(b' \,|\, u_{j_2-1}, u_{j_2}) = 1, \end{aligned}$$

hold. Thus we have

$$\begin{pmatrix} X'_{j_2-1} \\ Y'_{j_2-1} \end{pmatrix} \in \left\{ \begin{pmatrix} 1 \\ 1 \end{pmatrix}, \begin{pmatrix} 1 \\ 2 \end{pmatrix}, \begin{pmatrix} 2 \\ 2 \end{pmatrix}, \begin{pmatrix} 2 \\ 3 \end{pmatrix}, \ldots, \begin{pmatrix} b'-1 \\ b'-1 \end{pmatrix}, \begin{pmatrix} b'-1 \\ b' \end{pmatrix} \right\}.$$

If $X'_{j_2-1} = Y'_{j_2-1}$, the supporting pair of $\{X', Y'\}$ is $\{j_1, j_2\}$ and so $d(X', Y') = d(X, Y)$. If $X'_{j_2-1} \neq Y'_{j_2-1}$, the supporting pair of $\{X', Y'\}$ is $\{j_1, j_2 - 1\}$ and so $d(X', Y') = d(X, Y) - (n - j_2 + 1)/(n - 1)$.

Lemma 6 in Appendix section implies that if $u_{j_2-1} \geq u_{j_2}$, then

$$\begin{aligned} &\Pr[X'_{j_2-1} = Y'_{j_2-1} \mid \lfloor \Lambda \rfloor = j_2 - 1] \leq (1/2), \\ &\Pr[X'_{j_2-1} \neq Y'_{j_2-1} \mid \lfloor \Lambda \rfloor = j_2 - 1] \geq (1/2), \end{aligned}$$

by showing that the following inequality

$$\begin{aligned} \Pr[X'_{j_2-1} \neq Y'_{j_2-1} &\mid \lfloor \Lambda \rfloor = j_2 - 1] - \Pr[X'_{j_2-1} = Y'_{j_2-1} \mid \lfloor \Lambda \rfloor = j_2 - 1] \\ &= \sum_{k=1}^{b'-1} [g_{b'}(k \mid u_{j_2-1}, u_{j_2}) - g_{b'+1}(k \mid u_{j_2-1}, u_{j_2})] \\ &\quad - \sum_{k=1}^{b'-1} [g_{b'+1}(k \mid u_{j_2-1}, u_{j_2}) - g_{b'}(k - 1 \mid u_{j_2-1}, u_{j_2})] \geq 0 \end{aligned}$$

hold when $u_{j_2-1} \geq u_{j_2}$ (see Figure 3).

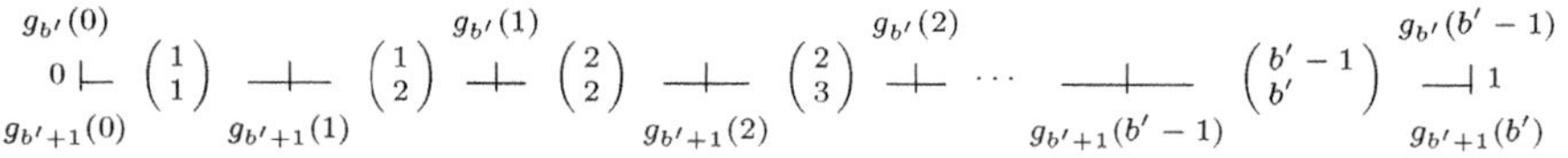

Fig. 3. A figure of alternating inequalities. In the above, we denote $g_{b'}(k \mid u_{j_2-1}, u_{j_2})$ and $g_{b'+1}(k \mid u_{j_2-1}, u_{j_2})$ by $g_{b'}(k)$, $g_{b'+1}(k)$, respectively.

Condition 1 is necessary to show Lemma 6. Thus we obtain that

$$\begin{aligned} &\mathrm{E}[d(X', Y') \mid \lfloor \Lambda \rfloor = j_2 - 1] \\ &\quad \leq (1/2) d(X, Y) + (1/2)\big(d(X, Y) - (n - j_2 + 1)/(n - 1)\big) \\ &\quad = d(X, Y) - (1/2)(n - j_2 + 1)/(n - 1). \end{aligned}$$

Case 2: When $\lfloor \Lambda \rfloor = j_2$, we can show that $\mathrm{E}[d(X', Y') \mid \lfloor \Lambda \rfloor = j_2] \leq d(X, Y) + (1/2)(n - j_2)/(n - 1)$ in a similar way with Case 1.

Case 3: When $\lfloor \Lambda \rfloor \neq j_2 - 1$ and $\lfloor \Lambda \rfloor \neq j_2$, it is easy to see that the supporting pair $\{j'_1, j'_2\}$ of $\{X', Y'\}$ satisfies $j_2 = \max\{j'_1, j'_2\}$. Thus $d(X, Y) = d(X', Y')$.

The probability of appearance of Case 1 is equal to $1/(n-1)$, and that of Case 2 is less than or equal to $1/(n-1)$. From the above,

$$\mathrm{E}[d(X', Y')] \leq d(X, Y) - \frac{1}{n-1} \frac{1}{2} \frac{n - j_2 + 1}{n - 1} + \frac{1}{n-1} \frac{1}{2} \frac{n - j_2}{n - 1}$$

$$
\begin{aligned}
&= d(X,Y) - \frac{1}{2(n-1)^2} \\
&\leq \left(1 - \frac{1}{2(n-1)^2}\frac{1}{\max_{\{X,Y\}\in\mathcal{E}}\{d(X,Y)\}}\right) d(X,Y) \\
&= \left(1 - \frac{1}{n(n-1)^2}\right) d(X,Y).
\end{aligned}
$$

Since the diameter of G is bounded by $n(\Delta - n)/2$, Theorem 5 implies that the mixing rate τ satisfies $\tau \leq n(n-1)^2\big(1+\ln(n(\Delta-n)/2)\big)$. □

Next, we estimate the coalescence time. When we apply Propp and Wilson's result in [15] straightforwardly, the obtained coalescence time is not tight. In the following, we use their technique carefully and derive a better bound.

Lemma 4. *Under Condition 1, the coalescence time* T_* *of* $\mathcal{M}$ *satisfies* $\mathrm{E}[T_*] = \mathrm{O}(n^3 \ln \Delta)$.

Proof. Let $G = (\Omega, \mathcal{E})$ be the undirected graph and $d(X,Y)$, $\forall X$, $\forall Y \in \Omega$, be the metric on G, both of which are defined in the proof of Lemma 3. We define $D \stackrel{\text{def.}}{=} d(X_{\mathrm{U}}, X_{\mathrm{L}})$ and $\tau_0 \stackrel{\text{def.}}{=} n(n-1)^2(1+\ln D)$. By using the inequality (1) obtained in the proof of Lemma 3, we have

$$
\begin{aligned}
\Pr[T_* > \tau_0] &= \Pr\big[\Phi^0_{-\tau_0}(X_{\mathrm{U}}, \boldsymbol{\Lambda}) \neq \Phi^0_{-\tau_0}(X_{\mathrm{L}}, \boldsymbol{\Lambda})\big] \\
&= \Pr\big[\Phi^{\tau_0}_0(X_{\mathrm{U}}, \boldsymbol{\Lambda}) \neq \Phi^{\tau_0}_0(X_{\mathrm{L}}, \boldsymbol{\Lambda})\big] \\
&\leq \sum_{(X,Y)\in\Omega^2} d(X,Y) \Pr\big[X = \Phi^{\tau_0}_0(X_{\mathrm{U}}, \boldsymbol{\Lambda}),\ Y = \Phi^{\tau_0}_0(X_{\mathrm{L}}, \boldsymbol{\Lambda})\big] \\
&= \mathrm{E}\big[d\big(\Phi^{\tau_0}_0(X_{\mathrm{U}}, \boldsymbol{\Lambda}), \Phi^{\tau_0}_0(X_{\mathrm{L}}, \boldsymbol{\Lambda})\big)\big] \leq \left(1 - \frac{1}{n(n-1)^2}\right)^{\tau_0} d(X_{\mathrm{U}}, X_{\mathrm{L}}) \\
&= \left(1 - \frac{1}{n(n-1)^2}\right)^{n(n-1)^2(1+\ln D)} D \leq \mathrm{e}^{-1}\mathrm{e}^{-\ln D} D \leq \frac{1}{\mathrm{e}}.
\end{aligned}
$$

By *submultiplicativity* of coalescence time ([15]), for any $k \in \mathrm{Z}_+$, $\Pr[T_* > k\tau_0] \leq (\Pr[T_* > \tau_0])^k \leq (1/\mathrm{e})^k$. Thus

$$
\begin{aligned}
\mathrm{E}[T_*] &= \sum_{t=0}^{\infty} t \Pr[T_* = t] \leq \tau_0 + \tau_0 \Pr[T_* > \tau_0] + \tau_0 \Pr[T_* > 2\tau_0] + \cdots \\
&\leq \tau_0 + \tau_0/\mathrm{e} + \tau_0/\mathrm{e}^2 + \cdots = \tau_0/(1 - 1/\mathrm{e}) \leq 2\tau_0.
\end{aligned}
$$

Clearly, $D \leq n(\Delta - n)/2 \leq \Delta^2$ because $n \leq \Delta$. Then we obtain the result that $\mathrm{E}[T_*] = \mathrm{O}(n^3 \ln \Delta)$. □

Lastly, we bound the expected number of transitions executed in Algorithm 1.

Proof of Theorem 4:

We denote T_* be the coalescence time of our chain. Clearly T_* is a random variable. Put $K = \lceil \log_2 T_* \rceil$. Algorithm 1 terminates when we set the starting time period $T = -2^K$ at $(K+1)$st iteration. Then the total number of simulated transitions is bounded by $2(2^0+2^1+2^2+\cdots+2^K) < 2\cdot 2\cdot 2^K \leq 8T_*$, since we need to execute two chains from both X_{U} and X_{L}. Thus the expectation of total number of transitions of $\mathcal{M}$ required in our algorithm is bounded by $\mathrm{O}(\mathrm{E}[8T_*]) = \mathrm{O}(n^3 \ln N)$. □

We can assume Condition 1 by sorting parameters in $\mathrm{O}(n \ln n)$ time.

6 Conclusion

In this paper, we proposed a monotone Markov chain whose stationary distribution is a discretized Dirichlet distribution. By employing Propp and Wilson's result, on monotone CFTP algorithm, we can construct a perfect sampling algorithm. We showed that our Markov chain is rapidly mixing by using path coupling method. Thus the rapidity implies that our perfect sampling algorithm is a polynomial time algorithm. The obtained time complexity does not depend on the magnitudes of parameters.

We can reduce the memory requirement by employing Wilson's read-once algorithm in [18].

Appendix

Lemma 5 is essentialy equivalent to the lemma appearing in Appendix section of the paper [12] by Matsui, Motoki and Kamatani which deals with an approximate sampler for discretized Dirichlet distribution.

Lemma 6 is a new result obtained in this paper.

Lemma 5.

$$\forall b \in \{2, 3, \ldots\},\ \forall u_i,\ \forall u_j \geq 0,\ \forall k \in \{1, 2, \ldots, b\},$$
$$g_{b+1}(k-1 \mid u_i, u_j) \leq g_b(k-1 \mid u_i, u_j) \leq g_{b+1}(k \mid u_i, u_j).$$

Proof. In the following, we show the second inequality. We can show the first inequality in a similar way.

We denote $C(u_i, u_j, b+1) = C_{b+1}$ and $C(u_i, u_j, b) = C_b$ for simplicity. From the definirion of $g_b(k \mid u_i, u_j)$, we obtain that

$$
\begin{aligned}
H(k) &\overset{\text{def.}}{=} g_{b+1}(k|u_i,u_j)-g_b(k-1|u_i,u_j)\\
&= \sum_{l=1}^{k} C_{b+1} l^{u_i-1}(b-l+1)^{u_j-1} - \sum_{l=1}^{k-1} C_b l^{u_i-1}(b-l)^{u_j-1}\\
&= \left(1-C_{b+1}\sum_{l=k+1}^{b} l^{u_i-1}(b-l+1)^{u_j-1}\right)-\left(1-C_b\sum_{l=k}^{b-1} l^{u_i-1}(b-l)^{u_j-1}\right)\\
&= C_b\sum_{l=k+1}^{b}(l-1)^{u_i-1}(b-l+1)^{u_j-1}-C_{b+1}\sum_{l=k+1}^{b} l^{u_i-1}(b-l+1)^{u_j-1}\\
&= \sum_{l=k+1}^{b}\left(C_b(l-1)^{u_i-1}(b-l+1)^{u_j-1}-C_{b+1}l^{u_i-1}(b-l+1)^{u_j-1}\right)\\
&= \sum_{l=k+1}^{b} C_b l^{u_i-1}(b-l+1)^{u_j-1}\left(\left(1-\frac{1}{l}\right)^{u_i-1}-\frac{C_{b+1}}{C_b}\right).
\end{aligned}
$$

Similarly, we can also show that

$$
\begin{aligned}
H(k) &= C_{b+1}\sum_{l=1}^{k} l^{u_i-1}(b-l+1)^{u_j-1} - C_b\sum_{l=1}^{k-1} l^{u_i-1}(b-l)^{u_j-1}\\
&\geq C_{b+1}\sum_{l=2}^{k} l^{u_i-1}(b-l+1)^{u_j-1} - C_b\sum_{l=2}^{k}(l-1)^{u_i-1}(b-l+1)^{u_j-1}\\
&= \sum_{l=2}^{k}\left(C_{b+1}l^{u_i-1}(b-l+1)^{u_j-1}-C_b(l-1)^{u_i-1}(b-l+1)^{u_j-1}\right)\\
&= \sum_{l=2}^{k} C_b l^{u_i-1}(b-l+1)^{u_j-1}\left(\frac{C_{b+1}}{C_b}-\left(1-\frac{1}{l}\right)^{u_i-1}\right).
\end{aligned}
$$

By introducing the function $h: \{2,3,\ldots,b\} \to \mathrm{R}$ defined by $h(l) \overset{\text{def.}}{=} \left(1-\frac{1}{l}\right)^{u_i-1} - \frac{C_{b+1}}{C_b}$, we have the following equality and inequality

$$
H(k) = \sum_{l=k+1}^{b} C_b l^{u_i-1}(b-l+1)^{u_j-1}h(l) \tag{2}
$$

$$
\geq -\sum_{l=2}^{k} C_b l^{u_i-1}(b-l+1)^{u_j-1}h(l). \tag{3}
$$

(a) Consider the case that $u_i \geq 1$.

Since $u_i - 1 \geq 0$, the function $h(l)$ is monotone non-decreasing. When $h(k) \geq 0$ holds, we have $0 \leq h(k) \leq h(k+1) \leq \cdots \leq h(b)$, and so (2) implies the non-negativity $H(k) \geq 0$. If $h(k) < 0$, then inequalities

$h(2) \leq h(3) \leq \cdots \leq h(k) < 0$ hold, and so (3) implies that $H(k) \geq -\sum_{l=2}^{k} C_b l^{u_i-1}(b-l+1)^{u_j-1}h(l) \geq 0$.

(b) Consider the case that $0 \leq u_i \leq 1$.

Since $u_i - 1 \leq 0$, the function $h(l)$ is monotone non-increasing. If the inequality $h(b) \geq 0$ hold, we have $h(2) \geq h(3) \geq \cdots \geq h(b) \geq 0$ and inequality (2) implies the non-negativity $H(k) \geq 0$. Thus, we only need to show that $h(b) = \left(\frac{b-1}{b}\right)^{u_i-1} - \frac{C_{b+1}}{C_b} \geq 0$.

In the rest of this proof, we substitute $u_{i'} - 1$ by $\alpha_{i'}$ for all i'. We define a function $H_0(b, \alpha_i, \alpha_j)$ by $H_0(b, \alpha_i, \alpha_j) \stackrel{\text{def.}}{=} (b-1)^{\alpha_i} C_{b+1}^{-1} - b^{\alpha_i} C_b^{-1}$. It is clear that if the condition $[-1 \leq \forall \alpha_i \leq 0,\ -1 \leq \forall \alpha_j,\ \forall b \in \{2, 3, 4, \ldots\},\ H_0(b, \alpha_i, \alpha_j) \geq 0]$ holds, we obatin the required result that $h(b) \geq 0$ for each $b \in \{2, 3, 4, \ldots\}$. Now we transform the function $H_0(b, \alpha_i, \alpha_j)$ and obtain another expression as follows;

$$
\begin{aligned}
H_0(b,\alpha_i,\alpha_j) &= (b-1)^{\alpha_i}\sum_{k=1}^{b} k^{\alpha_i}(b-k+1)^{\alpha_j} - b^{\alpha_i}\sum_{k=1}^{b-1} k^{\alpha_i}(b-k)^{\alpha_j} \\
&= \sum_{k=1}^{b}(b-1)^{\alpha_i}k^{\alpha_i}(b-k+1)^{\alpha_j}\frac{(b-k)+(k-1)}{b-1} - b^{\alpha_i}\sum_{k=1}^{b-1} k^{\alpha_i}(b-k)^{\alpha_j} \\
&= \sum_{k=1}^{b-1}\Bigg[(b-1)^{\alpha_i}k^{\alpha_i}(b-k+1)^{\alpha_j}\left(\frac{b-k}{b-1}\right) + (b-1)^{\alpha_i}(k+1)^{\alpha_i}(b-k)^{\alpha_j}\left(\frac{k}{b-1}\right) \\
&\qquad - b^{\alpha_i}k^{\alpha_i}(b-k)^{\alpha_j}\Bigg] \\
&= \sum_{k=1}^{b-1}\frac{(b-1)^{\alpha_i}k^{\alpha_i}(b-k)^{\alpha_j}}{b-1}\Bigg[\left(1+\frac{1}{b-k}\right)^{\alpha_j}(b-k) + \left(1+\frac{1}{k}\right)^{\alpha_i}k \\
&\qquad - \left(\frac{b}{b-1}\right)^{\alpha_i}(b-1)\Bigg].
\end{aligned}
$$

Then it is enough to show that the function

$$
H_1(b,\alpha_i,\alpha_j,k) \stackrel{\text{def.}}{=} \left(1+\frac{1}{b-k}\right)^{\alpha_j}(b-k) + \left(1+\frac{1}{k}\right)^{\alpha_i}k - \left(\frac{b}{b-1}\right)^{\alpha_i}(b-1)
$$

is nonnegative for any $k \in \{1, 2, \ldots, b-1\}$. Since $1 + 1/(b-k) > 1$ and $\alpha_j \geq -1$, we have

$$
H_1(b,\alpha_i,\alpha_j,k) \geq H_1(b,\alpha_i,-1,k) = \frac{(b-k)^2}{b-k+1} + \left(1+\frac{1}{k}\right)^{\alpha_i}k - \left(\frac{b}{b-1}\right)^{\alpha_i}(b-1).
$$

We differentiate the function H_1 by α_i, and obtain the following

$$
\begin{aligned}
\frac{\partial}{\partial\alpha_i}H_1(b,\alpha_i,-1,k) &= \left(1+\frac{1}{k}\right)^{\alpha_i}k\log\left(1+\frac{1}{k}\right) - \left(\frac{b}{b-1}\right)^{\alpha_i}(b-1)\log\left(\frac{b}{b-1}\right) \\
&= \left(1+\frac{1}{k}\right)^{\alpha_i}\log\left(1+\frac{1}{k}\right)^{k} - \left(1+\frac{1}{b-1}\right)^{\alpha_i}\log\left(1+\frac{1}{b-1}\right)^{(b-1)}.
\end{aligned}
$$

Since k, b is a pair of positive integers satisfying $1 \leq k \leq b-1$, the non-positivity of α_i implies $0 \leq (1+1/k)^{\alpha_i} \leq (1+1/(b-1))^{\alpha_i}$ and $0 \leq \log(1+1/k)^k \leq \log(1+1/(b-1))^{b-1}$. Thus the function $H_1(b, \alpha_i, -1, k)$ is monotone non-decreasing with respect to $\alpha_i \leq 0$. Thus we have

$$H_1(b, \alpha_i, -1, k) \geq H_1(b, 0, -1, k) = \frac{(b-k)^2}{b-k+1} + \left(1+\frac{1}{k}\right)^0 k - \left(\frac{b}{b-1}\right)^0 (b-1)$$
$$= \frac{(b-k)^2}{b-k+1} + k - b + 1$$
$$= \frac{(b-k)^2 + 1^2 - (b-k)^2}{b-k+1} = \frac{1}{b-k+1} \geq 0. \qquad \square$$

Lemma 6. $\forall b \in \{2, 3, \ldots\}$, $\forall u_i \geq \forall u_j$,

$$\sum_{k=1}^{b-1} [g_b(k \mid u_i, u_j) - g_{b+1}(k \mid u_i, u_j)] - \sum_{k=1}^{b-1} [g_{b+1}(k \mid u_i, u_j) - g_b(k-1 \mid u_i, u_j)] \geq 0.$$

Proof. We denote $C(u_i, u_j, b+1) = C_{b+1}$ and $C(u_i, u_j, b) = C_b$ for simplicity. It is not difficult to show the following equalities,

$$G \stackrel{\text{def.}}{=} \sum_{k=1}^{b-1} [g_b(k \mid u_i, u_j) - g_{b+1}(k \mid u_i, u_j)] - \sum_{k=1}^{b-1} [g_{b+1}(k \mid u_i, u_j) - g_b(k-1 \mid u_i, u_j)]$$
$$= \sum_{k=1}^{b-1} g_b(k \mid u_i, u_j) - \sum_{k=1}^{b-1} g_{b+1}(k \mid u_i, u_j)$$
$$- \sum_{k=1}^{b-1} g_{b+1}(k \mid u_i, u_j) + \sum_{k=1}^{b-1} g_b(k-1 \mid u_i, u_j)$$
$$= \sum_{k=1}^{b-1} g_b(k \mid u_i, u_j) - \sum_{k=1}^{b-1} g_{b+1}(k \mid u_i, u_j)$$
$$- \sum_{k=1}^{b-1} g_{b+1}(k \mid u_i, u_j) + \sum_{k=2}^{b-1} g_b(k-1 \mid u_i, u_j)$$
$$= \sum_{k=1}^{b-1} C_b \sum_{l=1}^{k} l^{u_i-1}(b-l)^{u_j-1} - \sum_{k=1}^{b-1} C_{b+1} \sum_{l=1}^{k} l^{u_i-1}(b-l+1)^{u_j-1}$$
$$- \sum_{k=1}^{b-1} C_{b+1} \sum_{l=1}^{k} l^{u_i-1}(b-l+1)^{u_j-1} + \sum_{k=2}^{b-1} C_b \sum_{l=1}^{k-1} l^{u_i-1}(b-l)^{u_j-1}$$
$$= C_b \sum_{l=1}^{b-1} (b-l) l^{u_i-1}(b-l)^{u_j-1} - C_{b+1} \sum_{l=1}^{b-1} (b-l) l^{u_i-1}(b-l+1)^{u_j-1}$$
$$- C_{b+1} \sum_{l=1}^{b-1} (b-l) l^{u_i-1}(b-l+1)^{u_j-1} + C_b \sum_{l=1}^{b-1} (b-l-1) l^{u_i-1}(b-l)^{u_j-1}$$

$$
\begin{aligned}
&= C_b \sum_{l=1}^{b-1} (2b-2l-1) l^{u_i-1} (b-l)^{u_j-1} - C_{b+1} \sum_{l=1}^{b-1} (2b-2l) l^{u_i-1} (b-l+1)^{u_j-1} \\
&= C_b C_{b+1} \left(C_{b+1}^{-1} \sum_{l=1}^{b-1} (2b-2l-1) l^{u_i-1} (b-l)^{u_j-1} \right. \\
&\qquad \left. - C_b^{-1} \sum_{l=1}^{b} (2b-2l) l^{u_i-1} (b-l+1)^{u_j-1} \right) \\
&= C_b C_{b+1} \left(\sum_{k=1}^{b} k^{u_i-1} (b-k+1)^{u_j-1} \sum_{l=1}^{b-1} (2b-2l-1) l^{u_i-1} (b-l)^{u_j-1} \right. \\
&\qquad \left. - \sum_{k=1}^{b-1} k^{u_i-1} (b-k)^{u_j-1} \sum_{l=1}^{b} (2b-2l) l^{u_i-1} (b-l+1)^{u_j-1} \right) \\
&= C_b C_{b+1} \left(\sum_{k=1}^{b} \sum_{l=1}^{b-1} (2b-2l-1)(kl)^{u_i-1} ((b-k+1)(b-l))^{u_j-1} \right. \\
&\qquad \left. - \sum_{k=1}^{b-1} \sum_{l=1}^{b} (2b-2l)(kl)^{u_i-1} ((b-k)(b-l+1))^{u_j-1} \right) \\
&= C_b C_{b+1} \left(\sum_{k=1}^{b} \sum_{l=1}^{b-1} (2b-2l-1)(kl)^{u_i-1} ((b-k+1)(b-l))^{u_j-1} \right. \\
&\qquad \left. - \sum_{k=1}^{b} \sum_{l=1}^{b-1} (2b-2k)(kl)^{u_i-1} ((b-l)(b-k+1))^{u_j-1} \right) \\
&= C_b C_{b+1} \sum_{k=1}^{b} \sum_{l=1}^{b-1} \left((2k-2l-1)(kl)^{u_i-1} ((b-k+1)(b-l))^{u_j-1} \right) \\
&= \frac{C_b C_{b+1}}{2} \left(\sum_{k=1}^{b} \sum_{l=1}^{b-1} (2k-2l-1)(kl)^{u_i-1} ((b-k+1)(b-l))^{u_j-1} \right. \\
&\qquad \left. + \sum_{k=1}^{b} \sum_{l=1}^{b-1} (2k-2l-1)(kl)^{u_i-1} ((b-k+1)(b-l))^{u_j-1} \right) \\
&= \frac{C_b C_{b+1}}{2} \left(\sum_{k=1}^{b} \sum_{l=1}^{b-1} (2k-2l-1)(kl)^{u_i-1} ((b-k+1)(b-l))^{u_j-1} \right. \\
&\quad + \sum_{k=1}^{b} \sum_{l=1}^{b-1} (2(b-k+1) - 2(b-l) - 1) \\
&\qquad \left. ((b-k+1)(b-l))^{u_i-1} ((b-(b-k+1)+1)(b-(b-l)))^{u_j-1} \right)
\end{aligned}
$$

$$= \frac{C_b C_{b+1}}{2} \left(\sum_{k=1}^{b} \sum_{l=1}^{b-1} (2k-2l-1)(kl)^{u_i-1}((b-k+1)(b-l))^{u_j-1} \right.$$

$$\left. - \sum_{k=1}^{b} \sum_{l=1}^{b-1} (2k-2l-1)((b-k+1)(b-l))^{u_i-1}(kl)^{u_j-1} \right)$$

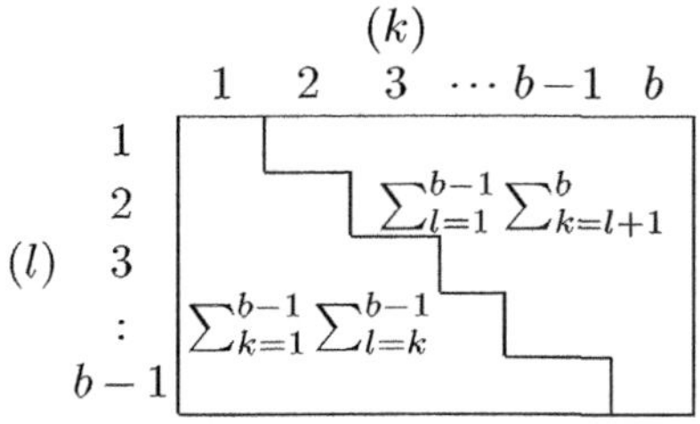

$$= \frac{C_b C_{b+1}}{2} \left(\sum_{l=1}^{b-1} \sum_{k=l+1}^{b} (2k-2l-1)(kl)^{u_i-1}((b-k+1)(b-l))^{u_j-1} \right.$$

$$+ \sum_{k=1}^{b-1} \sum_{l=k}^{b-1} (2k-2l-1)(kl)^{u_i-1}((b-k+1)(b-l))^{u_j-1}$$

$$- \sum_{l=1}^{b-1} \sum_{k=l+1}^{b} (2k-2l-1)((b-k+1)(b-l))^{u_i-1}(kl)^{u_j-1}$$

$$\left. - \sum_{k=1}^{b-1} \sum_{l=k}^{b-1} (2k-2l-1)((b-k+1)(b-l))^{u_i-1}(kl)^{u_j-1} \right)$$

$$= \frac{C_b C_{b+1}}{2} \left(\sum_{l=1}^{b-1} \sum_{k=l+1}^{b} (2k-2l-1)(kl)^{u_i-1}((b-k+1)(b-l))^{u_j-1} \right.$$

$$- \sum_{l=1}^{b-1} \sum_{k=l+1}^{b} (2k-2l-1)(l(k-1))^{u_i-1}((b-l+1)(b-k+1))^{u_j-1}$$

$$- \sum_{l=1}^{b-1} \sum_{k=l+1}^{b} (2k-2l-1)((b-k+1)(b-l))^{u_i-1}(kl)^{u_j-1}$$

$$\left. + \sum_{l=1}^{b-1} \sum_{k=l+1}^{b} (2k-2l-1)((b-l+1)(b-k+1))^{u_i-1}(l(k-1))^{u_j-1} \right)$$

$$= \frac{C_b C_{b+1}}{2} \sum_{l=1}^{b-1} \sum_{k=l+1}^{b} (2k-2l-1) \Big((kl)^{u_i-1}((b-k+1)(b-l))^{u_j-1}$$

$$- (l(k-1))^{u_i-1}((b-l+1)(b-k+1))^{u_j-1}$$

$$- ((b-k+1)(b-l))^{u_i-1}(kl)^{u_j-1}$$

$$+ ((b-l+1)(b-k+1))^{u_i-1}(l(k-1))^{u_j-1} \Big).$$

We define a function $G_0(k,l,u_i,u_j)$ by

$$G_0(k,l,u_i,u_j) \stackrel{\text{def.}}{=} \begin{pmatrix} (kl)^{u_i-1}((b-k+1)(b-l))^{u_j-1} \\ -(l(k-1))^{u_i-1}((b-l+1)(b-k+1))^{u_j-1} \\ -((b-k+1)(b-l))^{u_i-1}(kl)^{u_j-1} \\ +((b-l+1)(b-k+1))^{u_i-1}(l(k-1))^{u_j-1} \end{pmatrix}.$$

Since $1 \leq l < l+1 \leq k \leq b$, it is clear that $(2k-2l-1) > 0$. Thus, we only need to show that $\forall l \in \{1,2,\ldots,b-1\}$, $\forall k \in \{2,3,\ldots,b\}$, $\forall u_i \geq \forall u_j$, $G_0(l,k,u_i,u_j) \geq 0$. It is easy to see that

$$G_0(k,l,u_i,u_j) = \begin{pmatrix} (l(k-1))^{u_i-1}(b-k+1)^{u_j-1}\left(\left(1+\frac{1}{k-1}\right)^{u_i-1}(b-l)^{u_j-1}-(b-l+1)^{u_j-1}\right) \\ +((b-k+1)(b-l))^{u_i-1}l^{u_j-1}\left(-k^{u_j-1}+\left(1+\frac{1}{b-l}\right)^{u_i-1}(k-1)^{u_j-1}\right) \end{pmatrix}.$$

Then it is clear that $G_0(k,l,u_i,u_j)$ is non-decreasing with respect to u_i and so $G_0(k,l,u_i,u_j) \geq G_0(k,l,u_j,u_j)$. By substituting u_i by u_j in the definition of $G_0(k,l,u_i,u_j)$, it is easy to see that $G_0(k,l,u_j,u_j) = 0$. Thus we have the desired result. □

References

1. Bubley, R. and Dyer, M. (1997) Path coupling: A technique for proving rapid mixing in Markov chains, *38th Annual Symposium on Foundations of Computer Science*, IEEE, San Alimitos, 223–231.
2. Bubley, R. (2001) *Randomized Algorithms: Approximation, Generation, and Counting*, Springer-Verlag, New York.
3. Burr, T.L. (2000) Quasi-equilibrium theory for the distribution of rare alleles in a subdivided population: justification and implications, *Ther. Popul. Biol.*, **57** 297–306.
4. Burr, D., Doss, H., Cooke, G.E. and Goldschmidt-Clermont, P.J. (2003) A meta-analysis of studies on the association of the platelet PlA polymorphism of glycoprotein IIIa and risk of coronary heart disease, *Stat. Med.*, **22** 1741–1760.
5. Dimakos, X.K. (2001) A guide to exact simulation, *International Statistical Review*, **69** 27–48.
6. Durbin, R., Eddy, R., Krogh, A. and Mitchison, G. (1998) *Biological sequence analysis: probabilistic models of proteins and nucleic acids*, Cambridge Univ. Press.
7. Graham, J., Curran, J. and Weir, B.S. (2000) Conditional genotypic probabilities for microsatellite loci, *Genetics*, **155** 1973–1980.
8. Kijima, S. and Matsui, T. Polynomial time perfect sampling algorithm for two-rowed contingency tables, *Random Structures and Algorithms* (to appear).
9. Kitada, S., Hayashi, T. and Kishino, H. (2000) Empirical Bayes procedure for estimating genetic distance between populations and effective population size, *Genetics*, **156** 2063–2079.

10. Laval, G., SanCristobal, M. and Chevalet C. (2003) Maximum-likelihood and Markov chain Monte Carlo approaches to estimate inbreeding and effective size form allele frequency changes, *Genetics*, **164** 1189–1204.
11. Matsui, T., Motoki, M. and Kamatani, N. (2003) Polynomial time approximate sampler for discretized Dirichlet distribution, *14th ISAAC 2003, Kyoto, Japan*, LNCS, Springer-Verlag, **2906** 676–685.
12. Matsui, T., Motoki, M. and Kamatani, N. (2003) Polynomial time approximate sampler for discretized Dirichlet distribution, METR 2003-10, Mathematical Engineering Technical Reports, University of Tokyo (available from http://www.keisu.t.u-tokyo.ac.jp/Research/techrep.0.html)
13. Niu, T., Qin, Z.S., Xu, X. and Liu, J.S. (2002) Bayesian haplotype inference for multiple linked single-nucleotide polymorphisms, *Am. J. Hum. Genet.*, **70** 157–169.
14. Pritchard, J.K., Stephens, M. and Donnely, P. (2000) Inference of population structure using multilocus genotype data, *Genetics*, **155** 945–959.
15. Propp, J. and Wilson, D. (1996) Exact sampling with coupled Markov chains and applications to statistical mechanics, *Random Structures and Algorithms*, **9** 232–252.
16. Propp, J. and Wilson, D. (1998) How to get a perfectly random sample from a generic Markov chain and generate a random spanning tree of a directed graph, *J. Algorithms*, **27** 170–217.
17. Robert, C.P. (2001) *The Bayesian Choice*, Springer-Verlag, New York.
18. Wilson, D. (2000) How to couple from the past using a read-once source of randomness, *Random Structures and Algorithms*, **16** 85–113.

The Optimal Receiver in a Chip-Synchronous Direct-Sequence Spread-Spectrum Communication System

Nobuoki Eshima

Department of Information Science, Faculty of Medicine,
Oita University, Oita 879-5593, Japan
e-mail: eshima@med.oita-u.ac.jp

Summary. The Markov Spread-Spectrum (SS) code and the Bernoulli code make signals with different properties in Direct-Sequence Spread-Spectrum (DS/SS) communication systems. The present paper derives the optimal receiver by using the signal properties. First, considering properties of SS signal the asymptotic equivalence of the Bernoulli and the Markov SS signal is proven in code acquisition. Second, the decoder of SS signal is mathematically defined, and the optimal decoder is given from a statistical decision theory. It is shown that the usual matched-filter receiver is optimal for the Bernoulli SS signal in a chip-synchronous DS/SS communication.

Key words: Bit Error Rate; Bernoulli Code; Matched-Filter Receiver; Markov Code; DS/SS Communication System

1 Introduction

The Direct-Sequence Spread-Spectrum (DS/SS) Communication System was first studied by the U.S. Military in 1940s, and the purposes of establishing this communication system are to transmit information secretly, to be difficult to monitor the communication, and to defense against interferences of noises on a communication. Since 1950s the communication system has been investigated for commercial applications (Scholtz, 1982). In order to work the DS/SS communication system effectively, the code acquisition must be performed as fast as possible for a high quality communication, and decoding with a low bit error rate has to be made to the best of the ability. The usual receiver for SS signal is a matched-filter receiver (Polydoros & Weber, 1984A, B), and its performances are affected by properties of SS codes. Although users' codes produced with the Linear Feedback Sift Register (LFSR), such as the Gold Code, were used for DS/SS communication systems, the sequence of codes are not random and thus the interferences from other users' signals does not form a normal distribution (Kohda, 2002). This is a drawback of the code-producing method. A general stochastic coding method is based on the simple Markov chain $\{X_i\}$ on the state space $\Omega = \{-1, 1\}$ with $\Pr(X_i = 1) = \frac{1}{2}$ and the transition probability $\Pr(X_{i+1} = -1 \mid X_i = -1) = \Pr(X_{i+1} = 1 \mid X_i = 1) = \frac{1+\lambda}{2}$, where $-1 < \lambda < 1$. We refer the code with $\lambda = 0$ to as the Bernoulli code, and the other cases the Markov codes. In order to provide random sequences of codes, a random coding method with chaotic dynamics was proposed for the DS/SS communication (Kohda, et al. 1992). If the data timing is completely known in an asynchronous DS/SS communication, the Markov code with $\lambda = \sqrt{3} - 2$ is superior to the Bernoulli code in minimizing the multiple access interferences (Mazzini, Rovatti & Setti (1999) and Ling & Li (2000)). The variances of multiple access interference are given in the sense of code and data average by Kohda & Fujisaki (2000). These discussions are based on outputs received with a matched-filter receiver. In order to compare the Bernoulli and the Markov SS signal, it is indispensable to consider properties of outputs for spreading chip times, because the matched-filter outputs are produced with a linear statistic of basic outputs for N spreading chip times.

The studies cited above are based on SS signals received with the matched-filter receiver; however an important question arises as to whether the usual matched-filter receiver is optimal. The aim of the present paper is to give a statistical solution to the question. This paper consists of four sections in addition to this section. In Section 2, properties of SS signals in a chip-synchronous DS/SS system are discussed, and Section 3 reviews a matched-filter receiver system. It is shown that the SS signals produced by the Bernoulli code and the Markov code are essentially equivalent in code acquisition. In Section 4, the optimal decoding method is proposed, and the lower limit of bit error rate is given in DS/SS communications. Finally, Section 5 provides a discussion and a conclusion to this study.

2 Matched-Filter Receiver

In this section, received signal in a DS/SS communication with $J+1$ users and its properties are briefly reviewed. Let $\{d_p^{(j)}\}_{p=0,\pm1,\pm2,\ldots}$ be the data sequence of the j-th user, and let $\mathbf{X}^{(j)} = \{X_q^{(j)}\}_{q=0,\pm1,\pm2,\ldots}$ be the code sequence of user j with period $N = \frac{T}{T_C}$, where T is the data period and T_C the chip time for spreading signal; N is a spreading factor; and where $d_p^{(j)}, X_q^{(j)} \in \{-1, 1\}$. Then, the data and the SS code waveform of the j-th user at time t, $d^{(j)}(t)$ and $X^{(j)}(t)$, are respectively given by

$$d^{(j)}(t) = \sum_{p=-\infty}^{+\infty} d_p^{(j)} u\left(\frac{t}{T} - p\right) \quad \text{and} \quad X^{(j)}(t) = \sum_{q=-\infty}^{+\infty} X_q^{(j)} u\left(\frac{t}{T_C} - q\right),$$

where

$$u(t) = \begin{cases} 1 & (0 \leq t < 1) \\ 0 & (\text{otherwise}) \end{cases}.$$

From this, the SS signal of the j-th user is expressed as

$$s^{(j)}(t) = d^{(j)}(t)X^{(j)}(t).$$

In a DS/SS communication, the received signal at time t, $r(t)$, is the sum of $J+1$ users' signals, i.e.,

$$r(t) = \sum_{i=1}^{J+1} s^{(j)}(t - t^{(j)}) + n(t) \tag{2.1}$$

where $t^{(j)}$ is the propagation delay of the j-th user and $n(t)$ is the white noise. It is assumed that $t^{(j)}$ are identically, independently and uniformly distributed on integer values $\{0, 1, 2, \ldots, N-1\}$ in a chip-synchronous communication, i.e. $t^{(j)} \in \{0, 1, 2, \ldots, N-1\}$. For simplicity of the discussion $n(t)$ is set to be zero, and $T_C = 1$, i.e. $T = N$. For code acquisition of the j-th user, the matched-filter output is given by

$$Z_{pi}^{(j)} = \int_{pN+i}^{(p+1)N+i} r(t)X^{(j)}(t-i)\,dt \qquad (i = 0, 1, 2, \ldots, N-1;\ p = 0, 1, 2, \ldots). \tag{2.2}$$

For simplicity of the discussion, the data sequence is assumed to be the Bernoulli sequence with $E\big(d_p^{(j)}\big) = 0$. In order to discuss properties of received SS signals, let us set

$$r_i = \int_i^{i+1} r(t)\,dt \qquad (i = 0, 1, 2, 3, \ldots). \tag{2.3}$$

The above system makes a basic part of the present discussion. Then, from (2.2) we have

$$\begin{aligned} Z_{pi}^{(j)} &= \sum_{i=0}^{N-1} \int_{pN+i+k-1}^{(p+1)N+i+k} r(t)X^{(j)}(t-i)\,dt \\ &= \sum_{i=0}^{N-1} X_q^{(j)} r_{i+k} \quad (k = 0, 1, 2, 3, \ldots). \end{aligned} \tag{2.4}$$

The above output $Z_{pi}^{(j)}$ is a statistic of the original data $\{r_k, r_{k+1}, \ldots, r_{k+N-1}\}$. From this, the information of matched-filter output sequence $\{Z_{pi}^{(j)}\}$ is less than that of basic output sequence $\{r_i\}$. Hence, in order to construct an effective receiver system in the DS/SS communication system, it is indispensable to make a mathematical and statistical discussion based on the basic outputs in (2.3).

3 Properties of SS Signal

Let X_i, $i = 0, 1, 2, \ldots, N-1$, be a Markov chain on state space $\{-1, 1\}$, such that the initial distribution is $\Pr(X_1 = a) = 1/2$ for $a = -1, 1$, and the transition matrix:

$$P = \begin{pmatrix} \frac{1+\lambda}{2} & \frac{1-\lambda}{2} \\ \frac{1-\lambda}{2} & \frac{1+\lambda}{2} \end{pmatrix}$$

where $-1 < \lambda < 1$. It is assumed that users' code sequences are produced with the above Markov chain. In a chip-synchronous case, the propagation delay of the j-th user $t^{(j)}$ is uniformly distributed on $\{0, 1, 2, \ldots, N-1\}$. The received signal values (2.3), r_0, r_1, r_2, $\ldots$, are the sum of the $J+1$ users' SS signals. Then, by directly calculating we have the following theorem.

Theorem 3.1. *If users' data $\{d_p\}$ are Bernoulli trials with mean zero, for sufficiently large N the sequence of received signals has the following asymptotic covariances:*

$$\operatorname{Cov}(r_i, r_{i+k}) = \frac{(J+1)(N-k)\lambda^k}{N} \qquad \text{for } 0 \le k < N,$$

$$\text{and} \quad \operatorname{Cov}(r_i, r_{i+k}) = 0 \qquad \text{for } k \ge N. \quad \square$$

Proof. Since the users' codes are independently assigned to $J+1$ users, it is sufficient to consider the covariance of each user's outputs. Let $\{d_p\}_{p=0,\pm1,\pm2,\ldots}$ and $\{X_p\}_{p=0,\pm1,\pm2,\ldots}$ be the data and the code sequence of any user. Then, for $0 \le k < N$ we have

$$\begin{aligned}
&\operatorname{Cov}(r_i, r_{i+k}) \\
&= \operatorname{Cov}(r_0, r_{0+k}) \\
&= (J+1)\operatorname{Cov}(d_p X_q, d_{p'} X_{q+k}) \\
&= (J+1)\left(\frac{N-k}{N}\operatorname{Cov}(d_p X_q, d_p X_{q+k}) + \frac{k}{N}\operatorname{Cov}(d_p X_q, d_{p+1} X_{q+k})\right) \\
&= (J+1)\left(\frac{N-k}{N}\operatorname{Cov}(X_q, X_{q+k}) + \frac{k}{N}\operatorname{Cov}(d_p, d_{p+1})\operatorname{Cov}(X_q, X_{q+k})\right) \\
&= \frac{(J+1)(N-k)}{N}\operatorname{Cov}(X_q, X_{q+k}) \\
&= (J+1)(N-k)\frac{\lambda^k}{N}.
\end{aligned}$$

For $k \ge N$, we get

$$\operatorname{Cov}(r_i, r_{i+k}) = (J+1)\operatorname{Cov}(d_p X_q, d_{p+1} X_{q+k}) = 0.$$

This completes the theorem. $\square$

Since the spreading factor N is sufficiently large and $|\lambda|$ is less than 1, we can set $\lambda^k = 0$ for $|\lambda| < 0.5$ and $k > 5$. From this, for sufficiently large N we get

$$\mathrm{Cov}(r_i, r_{i+k}) = (J+1)\lambda^k \qquad (k = 0, 1, 2, \ldots)$$

The above result is derived from a general Markov code. Let $\mathbf{\Sigma}_{(\lambda)N}$ be the conditional covariance matrix of basic signal vector $\mathbf{r}_k = (r_k, r_{k+1}, \ldots, r_{k+N-1})$. Then, we have

$$\mathbf{\Sigma}_{(\lambda)N} = J \begin{pmatrix} 1 & \lambda & \lambda^2 & \ldots & \lambda^{N-1} \\ \lambda & 1 & \lambda & \ldots & \lambda^{N-2} \\ & & \ldots & & \\ \lambda^{N-2} & \lambda^{N-3} & \ldots & 1 & \lambda \\ \lambda^{N-1} & \lambda^{N-2} & \ldots & \lambda & 1 \end{pmatrix}. \tag{3.1}$$

The asymptotic distribution of the basic signal vector is the multivariate normal distribution with mean vector 0 and the above covariance matrix (3.1). Let us set

$$U_i = \begin{cases} r_0 & (i = 0) \\ \dfrac{r_i - \lambda r_{i-1}}{\sqrt{1-\lambda^2}} & (i > 1) \end{cases}. \tag{3.2}$$

Then, for integer values $i \geq 0$ we have

$$\begin{aligned} \mathrm{Cov}(U_i, U_{i+k}) &= \frac{1}{1-\lambda^2}\{\mathrm{Cov}(r_i, r_{i+k}) - \lambda\,\mathrm{Cov}(r_{i-1}, ri_{+k}) \\ &\quad - \lambda\,\mathrm{Cov}(r_i, r_{i+k-1}) + \lambda^2\,\mathrm{Cov}(r_{i-1}, ri_{+k-1)}\} \\ &= \begin{cases} J & (k = 0) \\ 0 & (k > 0) \end{cases} \end{aligned} \tag{3.3}$$

for integer values $i \geq 0$. Then, we have the following theorem.

Theorem 3.2. *For large integer J, outputs U_i $(i = 0, 1, 2, \ldots)$ in the sequence (3.3) are asymptotically independent and have the same asymptotic normal distribution with mean 0 and variance J.* □

For simplicity of the notation, let $(X_0, X_1, \ldots, X_{N-1})$ be the code of a target user, i.e. one of the $J+1$ users in the same system. Here, we consider the following receiver system.

$$\begin{aligned} Z_i &= \sum_{j=0}^{N-1} U_{i+j} \frac{X_j - \lambda X_{j-1}}{\sqrt{1-\lambda^2}} \\ &= \frac{1}{1-\lambda^2} \sum_{j=0}^{N-1} (r_{i+j} - \lambda r_{i+j-1})(X_j - \lambda X_{j-1}). \end{aligned} \tag{3.4}$$

If signal $r_{i+j} - \lambda r_{i+j-1}$ includes the user's code $X_j - \lambda X_{j-1}$, from the large sample theory the conditional expectation of the above output given the users code $\mathbf{X}$ is as follows: As N goes to infinity,

$$E\left(\frac{Z_i}{N}\right) = d\sum_{j=0}^{N-1} \frac{(X_j - \lambda X_{j-1})(X_j - \lambda X_{j-1})}{N(1-\lambda^2)}$$
$$= d\sum_{j=0}^{N-1} \frac{1 - \lambda X_j X_{j-1}}{N(1-\lambda^2)} \longrightarrow d \quad \text{(in probability)},$$

where $X_{-1} = X_{N-1}$ and d is a data in a data period. If for $k \geq 1$ and binary data d basic outputs r_{i+j} include X_{j+k}, $j = 0, 1, 2, \ldots, N-k$, and if for $k \geq 1$ and binary data d' basic outputs r_{i+j} include $d'X_{j+k-N}$ ($j = N-k+1, N-k+2, \ldots, N-1$), the correlator output is not synchronous and we have the following asymptotic conditional expectation. As N goes to infinity,

$$E(Z_i) = d\sum_{j=0}^{N-k-1} \frac{(X_{j+k} - \lambda X_{j+k-1})(X_j - \lambda X_{j-1})}{N(1-\lambda^2)}$$
$$+ d'\sum_{j=N-k}^{N-1} \frac{(X_{j+k} - \lambda X_{j+k-1})(X_j - \lambda X_{j-1})}{N(1-\lambda^2)} \longrightarrow 0 \quad \text{(in probability)}.$$

From this, for sufficiently large spreading factor N we can treat the asymptotic conditional expectations of Z_i as

$$E(Z_i) = \begin{cases} Nd & (\text{If } E(r_{i+j} - \lambda r_{i+j-1}) = d(X_j - \lambda X_{j-1})) \\ \lambda Nd & (\text{If } E(r_{i+j} - \lambda r_{i+j-1}) = d(X_{j-1} - \lambda X_{j-2})) \\ 0 & (\text{otherwise}) \end{cases}. \tag{3.5}$$

Remark 3.1. In (3.4), if for relatively small integer k (> 0) compared with spreading factor N,

$$E(r_{i+j} - \lambda r_{i+j-1}) = d(X_{j-k} - \lambda X_{j-k-1}),$$

then

$$E(Z_i) = \lambda^k Nd.$$

In this paper, for $k \geq 1$ the values $\lambda^k Nd$ are negligible compared with Nd, because $\lambda = \sqrt{3} - 2 = -0.268$. □

Since users' codes are independent, from (3.3) we have the asymptotic conditional covariances of Z_i as follows: As N goes to infinity,

$$\frac{\mathrm{Cov}(Z_i, Z_{i+k})}{N} \longrightarrow 0 \quad \text{(in probability)} \qquad \text{for integer value } k > 0,$$
$$\frac{\mathrm{Var}(Z_i)}{N} \longrightarrow J \quad \text{(in probability)}.$$

From this, for sufficiently large N we can set

$$\mathrm{Cov}(Z_i, Z_{i+k}) = 0 \text{ for integer value } k > 0; \text{ and } \mathrm{Var}(Z_i) = NJ.$$

From the above results, we have the following theorems.

Theorem 3.3. *For sufficiently large integer NJ, the sequence $\{Z_i\}$ in (3.4) is asymptotically independent.* □

For Bernoulli code, (3.4) becomes

$$E(Z_i) = \begin{cases} Nd & (\text{If } E(r_{i+j}) = dX_j) \\ 0 & (\text{otherwise}) \end{cases} .$$

From the above discussion, we obtain the following results. In a chip-synchronous DS/SS communication, the SS signal produced with the Markov-Coding method with any λ is asymptotically equivalent to that with the Bernoulli-Coding method. Hence, code acquisition performances with the Bernoulli code and the Markov code are almost the same.

4 The Optimal Decoding of SS Signal

In this section, it is assumed that for a user with code $\mathbf{X} = (X_0, X_1, \ldots, X_{N-1})$ the SS signal is synchronized. First, the decoding of the SS signal for the user is defined as follows.

Definition 4.1. *Let $\mathbf{r} = (r_0, r_1, \ldots, r_{N-1})^T$ be a synchronous set of basic outputs for a user with code $\mathbf{X}$. Then, the decoding is to give a value $\delta(\mathbf{r} \mid \mathbf{X}) \in \{-1, +1\}$, where $\delta(\mathbf{r} \mid \mathbf{X})$ is an appropriate function of $\mathbf{r}$, $\delta\colon R^N \longrightarrow \{-1, 1\}$.*

□

We refer a decoder with function $\delta(\mathbf{r} \mid \mathbf{X})$ to as decoder $\delta(\mathbf{r} \mid \mathbf{X})$. The bit error probability of the decoder is defined as follows:

Definition 4.2. *Let d be the binary data of a user with code $\mathbf{X}$, and let $\mathbf{r}$ be the corresponding synchronous signal set. Then, the bit error probability per the data d is defined by* $\Pr(\delta(\mathbf{r} \mid \mathbf{X})d = -1)$. □

The optimal (best) decoder is defined as follows:

Definition 4.3. *If decoder $\delta_0(\mathbf{r} \mid \mathbf{X})$ minimizes the bit error probability among all decoders $\delta(\mathbf{r} \mid \mathbf{X})$, i.e.*

$$\Pr(\delta_0(r \mid X)d = -1) = \min_{\delta} \Pr(\delta(r \mid X)d = -1),$$

then the decoder is optimal. □

We can derive the optimal decoder from a statistical decision theory. Let $Nor_N(\mathbf{X}, \mathbf{\Sigma})$ and $Nor_N(-\mathbf{X}, \mathbf{\Sigma})$ be N-dimensional normal distributions with mean vectors $\mathbf{X}$ and $-\mathbf{X}$ and covariance matrix $\mathbf{\Sigma}$, respectively. Then, if $d = 1$, the signal r came from normal distribution $Nor_N(\mathbf{X}, \mathbf{\Sigma})$; and if $d = -1$, $\mathbf{r}$ came from $Nor_N(-\mathbf{X}, \mathbf{\Sigma})$. However, the data d cannot be observed directly, so we must estimate the data. It is equivalent to decide the population from which the SS signal $\mathbf{r}$ came. The best way to decide the population of the signal is as follows (Anderson, 1983, Chapter 6, pp. 204–208):

$$\text{If } \mathbf{r}\boldsymbol{\Sigma}^{-1}\mathbf{X} > 0, \text{ then the signal is classified into } Nor_N(\mathbf{X}, \boldsymbol{\Sigma}); \text{ and otherwise, } Nor_N(-\mathbf{X}, \boldsymbol{\Sigma}). \tag{4.1}$$

The function of $\mathbf{r}$, $\mathbf{r}\boldsymbol{\Sigma}^{-1}\mathbf{X}$, is referred to as a linear discriminate function. In this decision procedure, the missclassification probability is given by

$$1 - \Phi\left(\frac{\Delta(\mathbf{X}, \boldsymbol{\Sigma})}{2}\right), \tag{4.2}$$

where

$$\Phi(y) = \int_{-\infty}^{y} \frac{1}{\sqrt{2\pi}} \exp\left(\frac{-x^2}{2}\right) dx,$$

and

$$\Delta(\mathbf{X}, \boldsymbol{\Sigma})^2 = 4 \operatorname{tr} \boldsymbol{\Sigma}^{-1}\mathbf{X}\mathbf{X}^T. \tag{4.3}$$

The quantity (4.3) is referred to as the Mahalanobis squared distance. The above decoder $\delta_0(r \mid X)$ (4.1) is optimal.

The above statistical discussion is applied in a chip-synchronous communication. The covariance matrix of $\mathbf{r} = (r_0, r_1, \ldots, r_{N-1})$ given a user's code $\mathbf{X}$ is given by $\boldsymbol{\Sigma}_{(\lambda)N}$ in (3.1). Before deriving the main results, we show the following lemmas.

Lemma 4.1. *Let*

$$\mathbf{P}_N = \begin{pmatrix} 1 & 0 & 0 & 0 & \ldots & 0 & 0 \\ \frac{-\lambda}{\sqrt{1-\lambda^2}} & \frac{1}{\sqrt{1-\lambda^2}} & 0 & 0 & \ldots & 0 & 0 \\ 0 & \frac{-\lambda}{\sqrt{1-\lambda^2}} & \frac{1}{\sqrt{1-\lambda^2}} & 0 & \ldots & 0 & 0 \\ \ldots & \ldots & \ldots & \ldots & \ldots & \ldots & \ldots \\ 0 & 0 & 0 & 0 & \ldots & \frac{1}{\sqrt{1-\lambda^2}} & 0 \\ 0 & 0 & 0 & 0 & \ldots & \frac{-\lambda}{\sqrt{1-\lambda^2}} & \frac{1}{\sqrt{1-\lambda^2}} \end{pmatrix}$$

Then,

$$\mathbf{P}_N \boldsymbol{\Sigma}_{(\lambda)N} \mathbf{P}_N^T = J\mathbf{I}_N,$$

where $\mathbf{I}_N$ is the $N \times N$ identity matrix.

Proof. By calculating $\mathbf{P}_N \boldsymbol{\Sigma}_{(\lambda)N} \mathbf{P}_N^T$, the theorem follows. □

By using the above lemma, we have the following lemma.

Lemma 4.2. *Let* $\Delta(\mathbf{X}, \mathbf{\Sigma}_{(\lambda)N})^2$ *be the Mahalanobis squared distance. Then, as spreading factor* N *goes infinity,*

$$\frac{J\Delta(\mathbf{X}, \mathbf{\Sigma}_{(\lambda)N})^2}{N} \longrightarrow 4 \qquad (\textit{in probability}).$$

Proof. From Lemma 4.1, we have

$$\begin{aligned}
J\Delta(\mathbf{\Sigma}_{(\lambda)N}, \mathbf{X})^2 &= 4J\,\mathrm{tr}\{(\mathbf{P}_N\mathbf{\Sigma}_{(\lambda)N}\mathbf{P}_N^T)^{-1}\mathbf{P}_N\mathbf{X}\mathbf{X}^T\mathbf{P}_N^T\} \\
&= 4\,\mathrm{tr}\,\mathbf{P}_N\mathbf{X}\mathbf{X}^T\mathbf{P}_N^T \qquad\qquad (\text{from Lemma 4.1}) \\
&= 4\Big\{X_1^2 + (N-1)\frac{1+\lambda^2}{1-\lambda^2} - \frac{2\lambda}{1-\lambda^2}\sum_{i=1}^{N-1} X_iX_{i+1}\Big\} \\
&= 4\Big\{1 + (N-1)\frac{1+\lambda^2}{1-\lambda^2} - \frac{2\lambda}{1-\lambda^2}\sum_{i=1}^{N-1} X_iX_{i+1}\Big\}.
\end{aligned}$$

As $N \longrightarrow \infty$,

$$\frac{\sum_{i=1}^{N-1} X_iX_{i+1}}{N} \longrightarrow \lambda \qquad (\text{in probability}).$$

This completes the lemma. □

From the above lemma, we obtain the following theorem.

Theorem 4.1. *For sufficiently large spreading factor* N *the asymptotic bit error probability of the optimal decoder is given by*

$$1 - \Phi\left(\sqrt{\frac{N}{J}}\right). \tag{4.4}$$

Proof. From Lemma 4.2, as $N \longrightarrow \infty$,

$$\frac{J}{N}\Delta(\mathbf{X}, \mathbf{\Sigma}_{(\lambda)N})^2 \longrightarrow 4 \qquad (\text{in probability}).$$

From (4.2), the theorem follows. □

From this it follows that *the bit error probability with the optimal decoding method is asymptotically equal to* (4.4) *for Markov SS signal.* The lower limit of the bit error probability for SS signals is given by (4.4), and the error probability is that of the usual matched-filter decoder for Bernoulli code signals.

5 Discussion

In this paper, we have proven almost the same performance of processing the Markov and the Bernoulli SS signal in a DS/SS communication, and have provided the optimal receiver for decoding SS signals. The present paper has shown that the usual matched-filter receiver is optimal for the Bernoulli SS signal in a chip-synchronous communication. Recently, it was shown that the Markov code with $\lambda = \sqrt{3} - 2$ is superior to the Bernoulli code with respect to bit error probability in decoding SS signals in an asynchronous DS/SS communication. A similar discussion can also be made for the asynchronous DS/SS communication.

In the DS/SS communication the optimal receiver can be constructed as in Section 4; however another question arises as to how a user's signal is completely synchronized, or synchronized as precisely as possible. For making the best performance in receiving SS signals, it is indispensable to use mathematical and statistical properties of the SS signals, and a theoretical discussion for it has to be made before establishing the receiver system.

Acknowledgement. This research was supported by Grant-in-aid for Scientific Research 16200021, Ministry of Education, Culture, Sports, Science and Technology of Japan.

References

1. Anderson, T.W. (1984) An Introduction to Multivariate Statistical analysis: CHAPTER 6, New York: John Wiley & Sons, Inc.
2. Kohda, T. (2002) Information sources using chaotic dynamics, *Proceedings of the IEEE*, **90**, 641–661.
3. Kohda, T. and Fujisaki, H. (2000) Variances of Multiple Access Interference: Code Average and Data Average, *Electronics Letters*, **36**, 20, 1717–1719.
4. Kohda, T., Tsuneda, A. and Sakae, T. (1992) Chaotic binary sequence by Chebychev maps and their correlation properties, in *Proceedings of IEEE 2nd International Symposium on Spread Spectrum Techniques and Applications*, 63–66.
5. Ling, G. and Li, S. (2000) Chaotic spreading sequences with multiple access performance better than random sequences, *IEEE Transactions on Circuits and Systems-I: Fundamental and Applications*, **47**, 394–397.
6. Mazzini, G., Rovatti, R. and Setti, G. (1999) Interference minimisation by auto-correlation shaping in asynchronous DS-CDMA systems: Chaos-based spreading is nearly optimal, *Electronics Letters*, **35**, 1054–1055.
7. Polydoros, A. and Weber, C.L. (1984A) A unified approach to serial search spread spectrum acquisition—Part I: General Theory, *IEEE Transactions on Communications*, **32**, 542–549.
8. Polydoros, A. and Weber, C.L. (1984B) A unified approach to serial search spread spectrum acquisition—Part II: A matched-filter receiver, *IEEE Transactions on Communications*, **32**, 550–560.

9. Scholtz, R.A. (1982) The origins of spread-spectrum communications, *IEEE Transactions on Communications*, **30**, 822–854.

Visualizing Similarity among Estimated Melody Sequences from Musical Audio

Hiroki Hashiguchi

Saitama University
hiro@ms.ics.saitama-u.ac.jp

Summary. We have developed a music retrieval system that receives a humming query and finds similar audio intervals (segments) in a musical audio database. This system enables a user to retrieve a segment of a desired musical audio signal just by singing its melody. In this paper, we propose a method to summarize the music database through similarity analysis to thereby reduce the retrieval time. The distance of chroma vectors is used as a similarity measure. The key technique for summarization includes, mainly, a statistical smoothing method and a method of discriminant analysis. Practical experiments were conducted using 115 musical audio selections in the RWC popular music database. We report the summarization ratio as about 45%.

Key words: Chroma vector, Discriminant analysis, Smoothing, Summarization of musical signals.

1 Introduction

Recently, music information resources, including audio CDs, MP3s, music titles, artists, have become globally ubiquitous. An industrial application to access a large amount of music information is an active research theme. Some activity of research on music information retrieval (MIR) can be found at the ISMIR homepage (http://www.ismir.net/). Music similarity analysis or pattern discovery for musical audio is a central theme of research of MIR, see Dannenberg and Hu [3].

From the viewpoint of summarization of music audio, Goto [6, 7, 8] proposed that a technique for detecting all similar segments (Refrain Detecting Method: RefraiD). Cooper and Foote [2] proposed a method to detect only a sabi section (a main part of a musical selection). Other methods for summarizing main parts and cutting the other parts were proposed in Dannenberg and Hu [3] and Peeters et al. [12].

* Contributed paper: received date: 8-Jul-05, accepted: 30-Sep-05

Statistical methods are gaining importance for finding similarity or repetition and to estimate melodies in music audio. Goto [4] estimated melodies by using EM algorithm.

In such a situation, Hashiguchi et al. [9] and Nishimura et al. [10] developed a music retrieval system using a humming query and proposed the retrieval techniques. This paper introduces methods to discover similar patterns of music audio and visualizes the estimated melody patterns. As described in Dannenberg and Hu [3], chroma vectors are helpful to discover the patterns. In this paper, we also compress similar intervals in a database to reduce the retrieval time of our music retrieval system.

2 Summarization of musical signals

The Key idea of our method in this section is based on the RefraiD method that was proposed by Goto [6, 7, 8]. The RefraiD method consists of stages of calculating chroma-based similarity in a time-lag plane, listing up repeated sections by using a smoothing technique and an automatic threshold selection technique based on a descriminant criterion, grouping the repeated sections, dealing with modulations (key changes) by shifting chroma vectors, and selecting the chorus sections. Our method was independently implemented by using different smoothing and grouping techniques that are simpler than the originals, and originally supports finding the end of summarization.

2.1 Definition of summarization

Typical popular music comprises, for example,

```
Intro → 1A Melody → 1B Melody → 1 sabi
      → 2A Melody → 2B Melody → 2 sabi → interval
                  → 3B Melody → 3 sabi
                              → 4 sabi → outro.
```

One of music's defining qualities is its refrain, as shown above. Similarity analysis intends to obtain the refrain by detecting similar melody intervals. We remake or cut the above melody sequence as follows

```
Intro → 1A Melody → 1B Melody → 1 sabi
```

and denote it here as "summarization" for musical audio.

2.2 Chroma similarity measure

We treat the range of music scales A1 (55 Hz) through A8 (7040 Hz). Let C denote the set of notes $C = \{1, \ldots, 12 \times 7\}$ whose element $c \in C$ corresponds to the frequency $\omega(c) = 55 \cdot 2^{(c-1)/12}$ [Hz]. We prepare WAV files in 16 kHz sampling and monaural recording as the music database. We adopt FFT analysis (2048 samples for each frame, 64 ms frame interval) and write the power spectrum as $f(w(c), t)$ for a frame t and a frequency $\omega(c)$. The total frame length of a song is denoted as W. For each note c ($1 \leq c \leq 12$), we define the additional power spectrum $v_c(t)$ as

$$v_c(t) = \sum_{h=1}^{7} f(\omega(12(h-1) + c), t). \tag{1}$$

Its vector $\mathbf{v}(t) = (v_1(t), \ldots, v_{12}(t))$ is called a "chroma vector". *Chroma similarity measure* $r(t, l)$ between $\mathbf{v}(t)$ and $\mathbf{v}(t-l)$, is defined as

$$r(t, l) = 1 - \frac{1}{\sqrt{12}} \left| \frac{\mathbf{v}(t)}{\max_c v_c(t)} - \frac{\mathbf{v}(t-l)}{\max_c v_c(t-l)} \right|. \tag{2}$$

Therein, l ($0 \leq l \leq t$) is called *lag*. The distance $|\cdot|$ takes an ordinary Euclid norm. The real number $\sqrt{12}$ in the denominator indicates the length of a diagonal line for a 12-dimensional unit hypercube and normalizes $r(t, l)$ to $0 \leq r(t, l) \leq 1$. Drawing the chroma similarity measure $r(t, l)$, as shown in Fig. 1 in the t-l plane with a frame t (the horizontal axis) and a lag l (the vertical axis), the lines of high similarity that are parallel to the t axis appear as bold traces in the lower triangular half of the plane. Originally, the definition (2) was adopted from Goto [6]. Another definition, by Bartsch and Wakefield [1], of the chroma similarity measure is that its distance is defined as the Euclidean distance between vectors normalized to have a mean of zero and a standard deviation of one.

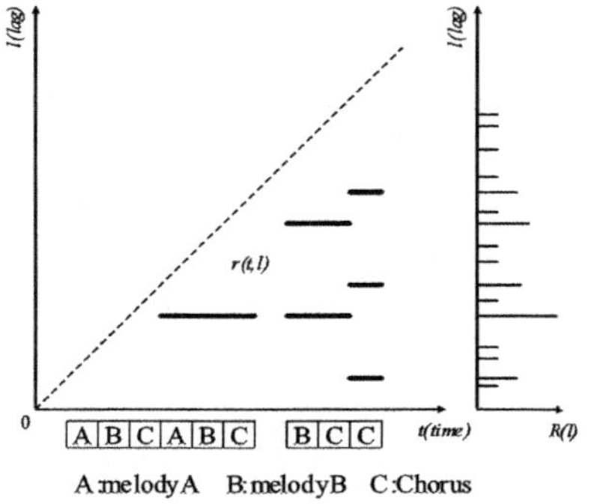

Fig. 1. Illustration of chroma's similarity measure: $r(t, l)$
The original figure of Fig. 1 is shown in Goto [6, 7, 8] with detail explanation.

The first figure in Fig. 2 shows a part of $r(t, l)$ from an actual musical audio selection. As described in Goto [6], the noises of $r(t, l)$ appear on the upper right, lower left, upward and downward. We carry out a method to reduce the noise in the same manner of Goto [6] and write it as $r_d(t, l)$ after application of the method. The center figure of Fig. 2 shows the values of $r_d(t,l)$ after reducing noise of the left.

(Left side) $r(l,t)$: chroma's similarity measure in part of an actual musical audio selection
(Center) $r_d(t,l)$: Noise reduction of $r(l,t)$ above
(Right side) $L(l)$: Lines corresponding to a similar segment

Fig. 2. Detection of similar segments

Illustrating $r_d(t,l)$ in t-l plane as Fig. 1, for each fixed l we can see faint similar lines:

$$\{(t,l) \,|\, t \in [t_1,t_2] \subset [0,W],\ r_d(t,l) \neq 0\}.$$

2.3 Detection of refrain

There are faint similar lines for $r_d(t,l) \neq 0$ almost everywhere in the t-l plane, as shown in the center of Fig 2. But we want to extract *similar lines* with height values of $r_d(t,l)$. We describe below how to detect the lag l with a high possibility of similar lines.

For any l, the peak $R(l)$, which is a moving average with $2Z_0+1$ frames $[t-Z_0, t+Z_0]$, is approximated by the equation (3). This $R(l)$ is illustrated in the right of Fig. 1.

$$R(l) = \sup_{l \le t \le W} \int_{t-Z_0}^{t+Z_0} \frac{r_d(\tau,l)}{2Z_0}\,d\tau \approx \max_{l \le t \le W} \frac{1}{2Z_0+1} \sum_{\tau=\max\{0,t-Z_0\}}^{\min\{t+Z_0,W\}} r_d(\tau,l) \tag{3}$$

Moreover, we take the moving average at $R(l)$ from $l-Z_1$ to $l+Z_1$ and subtract this moving average from $R(l)$ in order to reduce local variation at l. Here we write the value after this subtraction by $R'(l)$. This calculation is equivalent to the process of a high-pass filter.

$$R'(l) = \max\left\{0, R(l) - \int_{l-Z_1}^{l+Z_1} \frac{R(\xi)}{2Z_1}\,d\xi\right\} \tag{4}$$

$$\approx \max\left\{0, R(l) - \frac{1}{2Z_1+1}\sum_{\xi=\max\{0,l-Z_1\}}^{\min\{l+Z_1,W\}} R(\xi)\right\} \tag{5}$$

Making the distribution of $\{R'(1),\ldots,R'(W)\}$, discriminant analysis for two groups, whether the similarity line exists or not is carried out by the method of Ohtsu [11]. This method is the determination of a threshold α, so that the between-variance

$$(\lambda_1 - \lambda_2)^2 p_1 p_2$$

attains the maximum, where λ_i $(i = 1, 2)$ is the mean of $Class_i$ and p_i $(i = 1, 2)$ is the probability of $Class_i$:

$$p_1 = \frac{\#\{l \mid R'(l) > \alpha\}}{W}, \quad p_2 = 1 - p_1.$$

Next, for some l satisfying $R'(l) > \alpha$, we take the moving average $r_s(t,l)$ of $r_d(t,l)$'s on $[t - Z_2, t + Z_2]$. This approximation is given as follows.

$$r_s(t,l) = \int_{t-Z_2}^{t+Z_2} \frac{r_d(\tau,l)}{2Z_2}\, d\tau \approx \frac{1}{2Z_2+1}\sum_{\tau=\max\{0,t-Z_2\}}^{\min\{t+Z_2,W\}} r_d(\tau,l).$$

Conversely, for some lag l such that $R'(l) \le \alpha$, we set $r_s(t,l) = 0$. Moreover, let $L(l)$ denote the set of all intervals $[t_1, t_2]$ satisfying $t_2 - t_1 > Z_3$, in addition to $r_s(t,l) > \beta$ for $\forall t \in [t_1, t_2]$ where Z_3 is a constant.

$$L(l) = \big\{[t_1,t_2] \,\big|\, r_s(t,l) > \beta \text{ for } \forall t \in [t_1,t_2],\ t_2 - t_1 > Z_3,\ 0 \le t_1 < t_2 \le W\big\}. \tag{6}$$

Here, β is obtained by the method of Ohtsu [11] as well as the deviation of α. In practical application, we set $Z_0 = 10$ (about 1.3 s), $Z_1 = Z_2 = 5$ and $Z_3 = 68$ (about 4.5 s).

2.4 Grouping

The similar line $\{(t,l) \mid t \in [t_1,t_2] \in L(l)\}$ is sensitive for small variations such as l and t. Although the length of the similar line is longer than Z_4, it might be very short. Therefore, we consider the integration of short lines to be robust for the variation.

First, for $[t_{1,1}, t_{1,2}] \in L(l)$ and $[t_{2,1}, t_{2,2}] \in L(l+1)$ such that $[t_{1,1}, t_{1,2}] \cap [t_{2,1}, t_{2,2}] \ne \emptyset$, we integrate $L(l)$ into $L(l+1)$ as follows. Using the integrating term

$$[t_{3,1}, t_{3,2}], \quad \text{s.t. } t_{3,1} = \min\{t_{1,1}, t_{2,1}\},\ t_{3,2} = \max\{t_{1,2}, t_{2,2}\}, \tag{7}$$

we set

$$L(l) := L(l) - \{[t_{1,1}, t_{1,2}]\}, \quad L(l+1) := \big(L(l+1) - \{[t_{2,1}, t_{2,2}]\}\big) \cup \{[t_{3,1}, t_{3,2}]\}.$$

We increment l from one up to the end and thereby obtain longer intervals.

Next, we consider the integration of the intervals in $L(l)$ for a fixed l.

For $[t_{1,1}, t_{1,2}]$, $[t_{2,1}, t_{2,2}] \in L(l)$, if either of the following conditions 1 or 2 holds, we merge $[t_{1,1}, t_{1,2}]$, $[t_{2,1}, t_{2,2}] \in L(l)$ by the equation (7), and we set $L(l)$ as

$$L(l) := \big(L(l) - \{[t_{1,1}, t_{1,2}],\ [t_{2,1}, t_{2,2}]\}\big) \cup \{[t_{3,1}, t_{3,2}]\}.$$

Conditions:

1. $|t_{1,1} - t_{2,1}| \le \min\big[\gamma \max\{t_{1,2} - t_{1,1}, t_{2,2} - t_{2,1}\}, Z_4\big]$
2. $|t_{1,2} - t_{2,2}| \le \min\big[\gamma \max\{t_{1,2} - t_{1,1}, t_{2,2} - t_{2,1}\}, Z_4\big]$
   Here, γ indicates the ratio to the allowing parameter of integration, while Z_4 represents the control parameter, not to exceed the length by integration.

We continue the above integration process until neither condition 1 nor 2 holds. In practical experiments, we set $\gamma = 0.2$ and $Z_4 = 100\gamma = 20$. We infer the elements of $L(l)$ as the same melody patterns when the above process ceases.

2.5 Deletion of an outro and an interval

All elements in $L(l)$ can be considered to be the same melody pattern. If the pattern does not appear in $L(l)$ before a half of a song, then it is deleted from $L(l)$ because main melody parts would appear until a half. If set $L(l)$ consists of only one element, then this melody pattern might be an interval or outro of a song. For that reason, we delete it from $L(l)$ and set $L(l) = \emptyset$.

2.6 Finding the end frame for summarization

One way is that we take the start frame of the second melody pattern after all melody patterns appear just one time to find the end frame for summarization of a music audio. For all elements $\big[t^{(l)}_{1,1}, t^{(l)}_{1,2}\big], \ldots, \big[t^{(l)}_{p,1}, t^{(l)}_{p,2}\big] \in L(l)$, we presume that $t^{(l)}_{1,1} < t^{(l)}_{2,1} < \cdots < t^{(l)}_{p,1}$, without loss of generality. We seek the start frame and choose the the end frame for summarization as follows:

$$t' = \max_{\{l \,|\, L(l) \neq \emptyset\}} \big\{t^{(l)}_{1,2} \,\big|\, [t_{1,1}, t_{1,2}] \in L(l)\big\}, \quad E = \min_{\{l \,|\, L(l) \neq \emptyset\}} \big\{t^{(l)}_{2,1} \,\big|\, t_{2,1} - l > t'\big\}.$$

3 Experimental results

3.1 Visualizing estimated melody pattern

We apply our method in section 2 to real musical audio files that include a 16 kHz sampling and a monaural recording. These files are a part of *RWC music database* made by Goto et al. [5]. We show the results of summarization in Figs. 3, 4 and 5. The first layer in these are colored from human listening, for example, the first layer in Fig. 3 indicates that *Cho-Cho* consists of two parts that have the same melody pattern. The second and third layers in Fig. 3 correspond to results of similarity analysis. Furthermore, the vertical line from the bottom to the second layer indicates the end frame of summarization E in section 2.6.

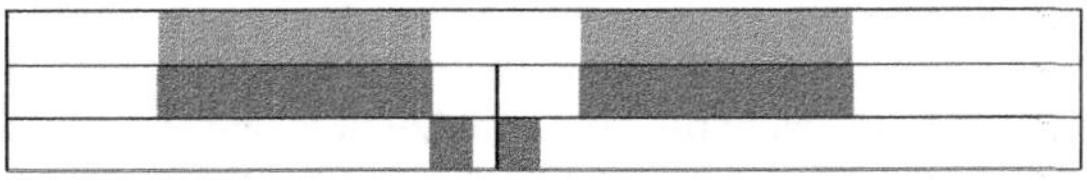

Fig. 3. RWC-MDB-R-2001 No. 1 Cho-Cho

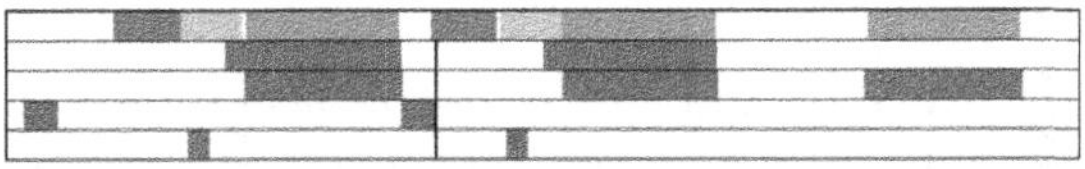

Fig. 4. RWC-MDB-P-2001 No. 5 Koi no Ver.2.4

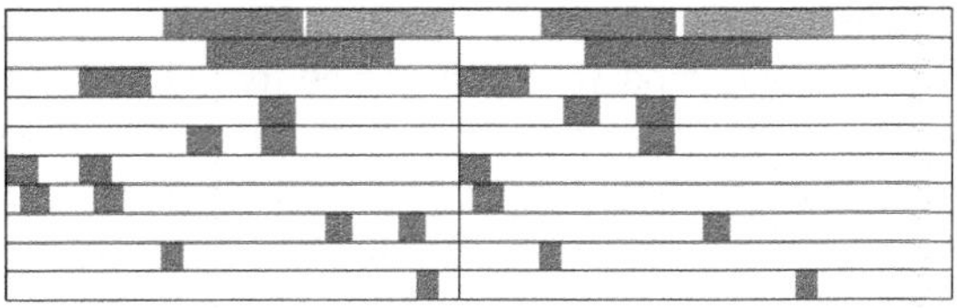

Fig. 5. RWC-MDB-P-2001 No. 12 KAGE-ROU

3.2 An application of similarity analysis to music retrieval by humming query

We conducted experiments on the summarization of a music database that consists of 100 popular music selections (RWC-MDB-P-2001 Nos. 1∼100) and 15 royalty-free music selections (RWC-MDB-R-2001 Nos. 1∼15).

The summarization ratio is about 45.3%, where it is defined as

$$\frac{\text{Total length after summarization}}{\text{Total length of the original}}.$$

The average summarization for all songs is 46.2%; the standard deviation is about 12.6%. The distribution for the ratios is shown in Fig. 6. The average of this ratio shows our methods might be conservative because any song has at least one refrain within its half.

Royalty-Free music selections consist of 15 *Doyo* (Children's Songs) whose construction might be simpler than that of popular music. The standard deviation of the Royalty-Free music category is 21.7% while the average is 51.4%. On the other hand, the standard deviation and the average for the popular music category is 10.7% and 44.9%, respectively. Therefore, the construction of a music song may cause the variety of the summarization.

The retrieval time for the database after summarization is about twice as fast as the original because the summarization ratio is about half. This efficiency was confirmed using our music retrieval system by humming query developed by Nishimura et al. [10].

4 Concluding remarks

This paper shows a method to summarize the music database through similarity analysis and an application to music retrieval system by humming query. The key technique for summarization includes, mainly, a statistical smoothing method and a method of discriminant analysis. As the most typical soothing method, a moving average is used for the smoothing of time series of chroma similarity. But other smoothing methods might be considerable and their comparison is one of feature works.

The author believes that this research is deeply relative to *KANSEI* information processing. At this viewpoint, statistical smoothing methods may extract rough trends that human beings feel similarity of melody, and discriminant analysis is used to determinate whether information on similarity is needed or not. The role of statistical methods in KANSEI information processing is our future work.

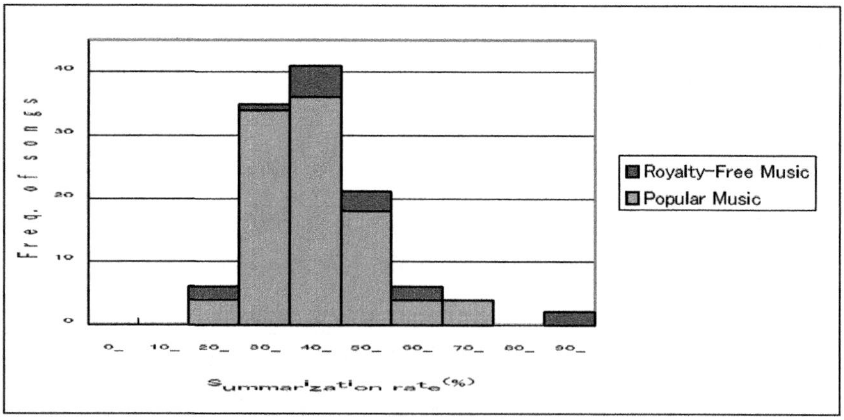

Fig. 6. Frequency of the summarization rate

Acknowledgement. I wish to thank two referees for their helpful comments on the paper.

This work was partially supported by Grant-in-Aid for Scientific Research (A)16200021 from Ministry of Education, Culture, Sports, Science and Technology and Research Project Fund A04-38 from Saitama University.

References

1. Bartsch, M. and Wakefield, G.H. (2001) To Catch a Chorus: Using Chroma-Based Representations For Audio Thumbnailing. in *Proceedings, of the Workshop on Applications of Signal Processing to Audio and Acoustics*, IEEE.
2. Cooper, M. and Foote, J. (2002) Automatic music summarization via similarity analysis, Proc. ISMIR, 81–85.
3. Dannenberg, R.B. and Hu, N. (2002) Pattern Discovery Techniques for Music Audio, Proc. ISMIR 2002, 63–70.
4. Goto, M. (2001) A Predominant-F0 Estimation Method for CD Recordings: MAP Estimation using EM Algorithm for Adaptive Tone Models, Proc. ICASSP 2001, V-3365–3368.
5. Goto, M., Hashiguchi, H., Nishimura, T. and Oka, R. (2002) RWC Music Database: Popular, Classical, and Jazz Music Databases, Proc. ISMIR 2002, 287–288.
6. Goto, M. (2002) A Real-time Music Scene Description System: A Chorus-Section Detecting Method, 2002-MUS-47-6, 2002 (100), 27–34 (in Japanese).
7. Goto, M. (2003) A Chorus-Section Detecting Method for Musical Audio Signals, Proceedings of the 2003 IEEE International Conference on Acoustics, Speech, and Signal Processing (ICASSP 2003), pp. V-437–440.
8. Goto, M. (2003) SmartMusicKIOSK: Music Listening Station with Chorus-Search Function, Proceedings of 16th Annual ACM symposium on User Interface Software and Technology (UIST 2003), pp. 31–40.
9. Hashiguchi, H., Nishimura, T., Takita, T., Zhang, J.X. and Oka, R. (2001) Music Signal Spotting Retrieval by a Humming Query, Proceedings of Fifth World Multi-Conference on Systemics, Cybernetics and Informatics, VII 280–284.
10. Nishimura, T., Hashiguchi, H., Takita, J., Zhang, J.X. and Oka, R. (2001) Music Signal Spotting Retrieval by a Humming Query Using Start Frame Feature Dependent Continuous Dynamic Programming, Proc. ISMIR 2001, 211–218.
11. Ohtsu. N. (1979) A threshold selection method from gray-level histograms, IEEE Trans. SMC, SMC-9 (1), 62–66.
12. Peeters, G., Burthe, A.L. and Rodet, X. (2002) Toward automatic music audio summary generation from signal analysis, Proc. ISMIR 2002, 94–100.

GPSR Compliance

The European Union's (EU) General Product Safety Regulation (GPSR) is a set of rules that requires consumer products to be safe and our obligations to ensure this.

If you have any concerns about our products, you can contact us on ProductSafety@springernature.com

In case Publisher is established outside the EU, the EU authorized representative is:

Springer Nature Customer Service Center GmbH
Europaplatz 3
69115 Heidelberg, Germany

Batch number: 10755053

Printed by Printforce, the Netherlands